MW01632624

SUPER CARS

This edition published by Barnes & Noble, Inc.,
by arrangement with Book Sales, Inc.

2004 Barnes & Noble Books

This edition produced for sales in the U.S.A., its
territories, and dependencies only.

M 10 9 8 7 6 5 4 3 2 1

ISBN 0-7607-6228-7

This book was designed and produced by
Quintet Publishing Limited
6 Blundell Street
London N7 9BH

Designed and Edited by Q2A Solutions

Associate Publisher: Laura Price
Creative Director: Richard Dewing

Art Director: Roland Codd
Project Editors: Jenny Doubt, Catherine Osborne

Manufactured in Singapore by Pica Digital Pte Ltd
Printed in Singapore by Star Standard Industries Pte Ltd.

The material used in this publication previously appeared in
Dream Cars by Mark Chivers et al. (ed.), *For Millionaires Only*
by Gordon Cruickshank, *Japanese Super Cars* by Terry Jackson,
and *Open Top Style* by Graham Robson.

SUPER CARS

CLASSICS OF THEIR TIME

BARNES & NOBLE BOOKS
NEW YORK

Contents

Introduction

In the beginning, cars were sometimes known as "horseless-carriages," as all of them were open-top types of one sort or another. Quite clearly, their styles had evolved from the coachwork of the horse-drawn carriages of an earlier era.

In later years, as tastes changed and the coachbuilder's art was refined to take account of this, more and more closed cars were sold.

A fine car has been an obvious expression of its owner's status and wealth from Edwardian days, when Mercedes and Rolls-Royce showed that the "horseless carriage" had developed into a refined and desirable means of transport.

Since those early years, from being the sole preserve of the wealthy, car ownership has gradually become more affordable to more people, thanks to the "people's cars" produced by Ford in the United States, Fiat in Italy, Citroën in France, and Austin and Morris in Great Britain. Today practically every family owns a car that is a technically better vehicle than the luxury limousine of the 1930s.

The huge gap between an ill-equipped, 30 mph (48 kph) light car and an 80 mph (129 kph) limousine boasting brocade and sterling silver trimmings, has vanished. Any car in

today's showrooms will cruise at 80 mph (129 kph) for hours, stay dry and warm in bad weather, and run for thousands of miles without attention. But it was for such technical qualities, which we now take for granted, that the wealthy bought Rolls-Royces or Hispano-Suizas before the war.

However, standards are still increasing, and quality still takes time and effort to ensure.

Fine cars also provide the intoxicating thrill of being at the wheel, and so offer a combination of both physical and visual pleasure.

An optimist might say that the range and availability of exciting cars is greater now than ever before. Whether we will be able to continue enjoying them in the years to come is hard to predict, considering the environmental problems that face us and diminishing fuel supplies.

Above **In the late 1930s, the BMW 328 sports car was one of the most desirable open-top cars in the world. Among its advanced features were a powerful 6-cylinder engine and independent front suspension.**

As the price of gas spirals, running a car may become an elite pastime once again. It may not be the cost of buying a car, but the expense of feeding it with gas that will make motoring a luxury.

A–Z of Super Cars

AC Cobra

V8 | 425 bhp | 160 mph (257 kph) | 1960s

The AC Cobra is considered the epitome of the 1960's sports car. It was an Anglo-American project, and the muscular machine, with its ear-splitting V8 wail, has become a legend on both sides of the Atlantic.

But when it was announced in 1962, many observers sneered at it, claiming that it was merely a hybrid. American race driver Carroll Shelby was behind the project, and his company turned the Cobra into a successful racer. A later coil-sprung, 7-liter version even won the World GT Championship in 1965. Demand for the car overtook the supply and production was cut off in 1968.

Below **Born again: brutal, loud, and fast, the AC Cobra of the 1960s has reappeared, mildly refined, as the AC MkIV. It's still loud and fast, but not quite so brutal.**

Alfa Romeo 2900B

Tipo B | 180 bhp | 140 mph (225 kph) | 1930s

Fewer than forty 2900Bs rolled out of the Alfa Romeo factory, which makes them one of the rarest and most sought-after classic cars.

Alfa's stock of Tipo B engines, in enlarged 2.9-liter twin-supercharger form, provided an ideal power source, and the factory drew up a sophisticated new chassis to utilize it. The car boasted an all-independent suspension, using twin trailing arms at the front and swing-axles with a transverse leaf spring at the tail.

Above **In the face of the all-conquering German steamroller in Grands Prix, Alfa Romeo switched its efforts to sports car racing, resulting in the sensational and advanced 2900B.**

A 4-speed gearbox in unit with the differential gave it excellent balance, especially in the shorter of its two wheelbases. Its acceleration was astonishing, and the supple suspension gave it remarkable road-holding.

Alfa Romeo 8C Monza

8-cylinder | 150 bhp (180 bhp for racing) | 135 mph (217 kph) | 1930s

The 8C series was Alfa's sporting mainstay from its introduction as a racer in 1931 until the enforced end of production in 1939. The car was designed by the greatest of all Alfa Romeo engineers – Vittorio Jano.

The 8C first appeared in a long wheelbase 2,300 sports car format. In the same year, it won the grueling Targa Florio race in Sicily and the Italian Grand Prix at Monza. From then on, the rareshort-chassis racing model has been known as the Monza.

At the heart of the 8C is Jano's magnificent, supercharged, twin-overhead-cam, straight-eight engine. Although the Alfa 8C started life as a 2.3-liter unit, it later grew to 2.6-liters and can reach 60 mph (97 kph) in less than 7 sec.

The car's controls are surprisingly modern in layout – the only exception being the placing of the throttle pedal between clutch and brakes. Its quick, accurate steering and powerful, mechanically operated brakes made it a truly delightful car to drive.

Below **Like many cars of the period, the Alfa Romeo Monza is supercharged to give it truly remarkable acceleration and a top speed of 135 mph (216 kph). The comfortable, well laid-out cockpit is relatively roomy, and has surprisingly comprehensive instrumentation.**

Alfa Romeo Giulia Spider

4-cylinder | 89 – 122 bhp | 105 – 120 mph | 1960 to present

In the 1950s, Alfa Romeo launched a series of smaller engined cars, the first of which was the Giulietta with an advanced 1.3 liter, 4-cylinder engine. Alfa then introduced the first of the four-door unit-body Giulia saloons in 1962, using a 1.6-liter version of the Giulietta's engine. The first open-top Giulia, called the 1600 Spider, appeared in 1966.

In 1969 the Spider was re-styled, with a sharply cut-off and shortened tail style, plus a more modern fascia, while the engine was enlarged to 1.8-liters.

Above **The early Spiders were called "Duetto," and had long and rounded tails, but 1969 Pininfarina modified the design by cutting the tail short. This was a 1978 model. The long sloping nose and the distinctively sculptured sides of this design remained the same, for more than 20 years.**

In the next few years 1.3-liter and 2-liter versions of the same engine were introduced. Then, in 1986, the car got a substantial facelift, complete with new front and rear spoilers, and skirts along the flanks.

Alfa Romeo GTV-6

V6 | 160 bhp | 127 mph (204 kph) | 1980s

One of the oldest Italian car makers, Alfa Romeo, has a marvelous background in making high-performance cars. One of them, the GTV-6, followed on from a long line of very fine 2+2-seater GT coupés that go back to the superb little Bertone-styled Giulietta of the early 1950s.

The three-door hatchback GTV-6 was powered by a 152 ci (2,492 cc) all-alloy V6 with twin overhead

camshafts, producing 160 bhp at 5,600 rpm. Bosch L-Jetronic fuel injection was used, together with electronic ignition. A 5-speed manual gearbox was mounted at the opposite end of the car, in unit with the differential.

The GTV-6's suspension was independent at the front, with torsion bars supplying the springing medium,

Left and Above **The Alfa Romeo GTV-6 was a car of real character, for better or worse. The attractive coupé had a fabulous V6 engine, exceptional roadholding and handling, a dreadful gearchange, and a typically Italian driving position — best suited to long arms and short legs. All the GTV-6's shortcomings, however, were trivial alongside its strengths.**

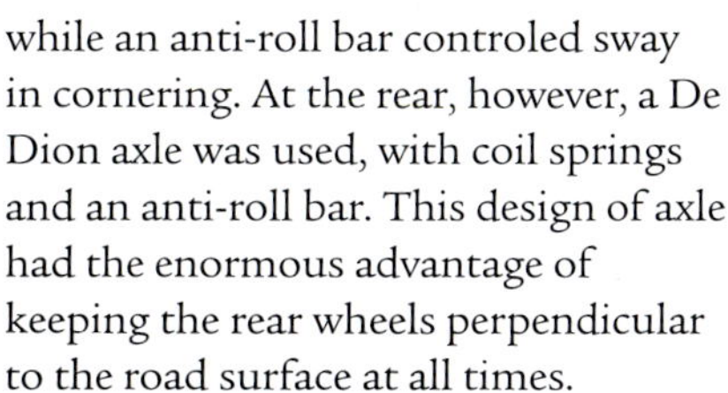

while an anti-roll bar controled sway in cornering. At the rear, however, a De Dion axle was used, with coil springs and an anti-roll bar. This design of axle had the enormous advantage of keeping the rear wheels perpendicular to the road surface at all times.

Brakes were disk all round, the fronts had ventilated rotors. The wheels were 15 x 6 in. (38 x 15 cm) light alloy components and the tires were 195/60HR 15. The car's balance was achieved by having the transmission located at the rear of this front-engined car.

Allard J2

V8 | 150 bhp | 120 mph (193 kph) | 1950s

Had Sydney Herbert Allard been alive today he would probably still be looking for the biggest, most powerful V8 he could find, and then building the smallest possible amount of car to put it in. Sophistication certainly was not Allard's forté, but he has to be respected for his unswerving conviction that anything, however unlikely, that made an Allard go faster, was good.

Surely the Allard that everyone remembers is the spectacular J2, a car that was equally successful as a road car and a racer.

Allard competed with the first J2 at the famous Prescott hill climb in the summer of 1949, winning his class in his 4.4-liter flathead Mercury V8 powered car. But the real performance was reserved for the United States, where the choice of engine was almost boundless, unlike in Britain, where the J2 was restricted to flathead engines.

With overhead-valve Ford, Chrysler, and Cadillac V8s, the J2 was a major success in America from its arrival in April 1950. Allard shipped rolling chassis with their good-looking, cycle-winged aluminum bodies and left the customer to drop in whatever engine would fit.

The chassis could certainly handle it, with deep, stiff side rails and ample cross-bracing. The front axle was a traditional Allard split-beam with coil springs, and the rear used a De Dion layout. The brakes were very good for their day, with hydraulically operated drums all around.

Allard's way of ensuring enough traction for the more potent J2s included hanging a 40 gallon fuel tank over the back axle. With massive V8 torque in a car weighing little more than a ton, it goes without saying that acceleration was spectacular even by today's standards.

Below **On the road or the racetrack, big is beautiful as far as Allards are concerned. This J2 model is powered by a 4.4-liter Mercury V8 engine – getting the power down onto the road is helped by slinging a large fueltank over the rear wheels.**

Aston Martin GT

V8 | N/A | N/A | 1960s

The Zagato-bodied lightweight DB4GT, first seen in 1960, represents the culmination of Aston Martin's efforts in the GT class. It is stunning to look at, muscular and compact, and possibly one of the greatest designs to emerge from the idiosyncratic house of Zagato.

There are 19 left of this model in total, all surviving and well documented. They vary slightly, as handmade cars will, but the most perfect in line and history are still known by their registration numbers, VEV 1 and VEV 2.

Above **Some race drivers claimed that the Zagato was no better than the plain GT, only slightly lighter, but less predictable.**

These two were made famous by Jim Clark, Roy Salvadori, and Innes Ireland. When both VEV 1 and VEV 2 came up for sale at separate auctions, collectors jumped at the idea of a matched pair. With a final combined price of almost 5.4USD (3GBP) million, the idea was then considered too much for any single buyer, and they went to different collectors.

Aston Martin Lagonda

V8 | N/A | 143 mph (230 kph) | 1980s

The Lagonda exuded luxury and high performance. It has a 326 ci (5,340 cc) double overhead cam V8 engine breathing through four Weber 42DCNF90/100 carburetors. Transmission from the front-mounted engine is through a Chrysler Torqueflite automatic gearbox to a chassis-mounted limited-slip differential.

Front suspension is by the classic unequal length wishbone, coil-spring, and damper, set up with an anti-roll bar. The rear suspension is self-leveling and features a De Dion axle.

Steering is by power-assisted rack and pinion, and brakes are ventilated disk units all round. The Lagonda uses 15 x 6 in. (38 x 15 cm) alloy wheels fitted as standard with the superb Avon Turbo speed 235/70 VR 15 tires.

The car's instruments used digital LED displays and covered all possible information that the driver needed. The Lagonda weighs 4,622 lb, has a top speed of 143 mph (230 kph).

Below **The 1980's Lagonda was considered one of the fastest cars in the world.**

Aston Martin Vantage

V8 | N/A | 168 mph (270 kph)| 1978 – 80s

The Aston Martin Vantage was a sports coupé in the old style – big, very fast, expensive, constructed with the finest materials, and beautifully finished.

The car used the same hand-built V8 engine as the Lagonda, but with high-performance camshafts and bigger Weber 481DF 3/150 carburetors increasing both power and torque. The gearbox was the ZF 5-speed manual model which was heavy in operation but otherwise superb. Wheel sizes were also better compared to the Lagonda's 15 x 8 in. (38 x 20 cm) size, and standard tires were 275/55 VR 15 Pirelli P7s.

The Vantage was shatteringly fast for its era, and could accelerate from 0 to 100 mph (0 to 161 kph) in 11.9 sec.

For such a large car, the Vantage was, surprisingly, a two-seater. Agility on twisty roads was further limited by its size, the car being much more suitable for long high-speed journeys, preferably by unrestricted motorway.

***Far Right* The Aston Martin Vantage and *Below* the drophead Volante both looked, and indeed felt, big, but their lusty V8 engines – all built by hand and each bearing its builder's name on a brass plate – made them two of the fastest cars in the world at the time when they were produced.**

Aston Martin Volante

V8 | 309 bhp | 145 mph (233 kph) | 1978 onward

Aston Martin, like other specialist car makers, always liked to have a convertible derivative of all its cars. In 1978, 11 years after the coupé had first appeared, the company offered the V8-engined Volante model.

The V8-engined car was based on a solidly engineered steel platform chassis which, because it was effectively hand-built, featured many small pressings or fabrications.

Much of the body shell was steel, but many skin panels were in aluminum alloy, the car body being an intriguing blend of machine-made pressings, hand-dressed, matched, and welded together at the Aston Martin factory at Newport Pagnell. Because it was a large and wide car, the coupé was a full four-seater.

The V8 Volante had the same basic style as the coupé, but naturally featured a power operated soft-top, and had its floorpan stiffened to restore the rigidity of the chassis. The fold-top had, however, encroached on the boot space (reduced from 8.6 cu ft to 5.1 cu ft).

The soft-top was beautifully tailored, and when raised, turned the Volante into a snug, if somewhat claustrophobic, two-door sedan.

At first the Volante was only available with the standard engine, but from 1986 an extra version, complete with deep front spoiler, was offered with the Vantage engine.

Right **The Aston Martin Volante – wood, British leather, a padded steering wheel, and an excellent speedometer reading.**

Below **With either the soft-top up or down, the Aston Martin Volante offered an impressive style. The later Volantes were also offered with the most powerful Vantage engine.**

Aston Martin Zagato

V8 | 432 bhp | 187 mph (301 kph) | 1980s

Inspired by the gorgeous lightweight DB4GT, during the 1980s, the Newport Pagnell firm decided to renew its successful association with Italian coachbuilder Zagato. The result was 50 Aston Martin Zagotas, a car that was most immediatly notable for the absolute minimal discomfort it offered its lucky owners.

The stripped-down and lightened floorpan of a V8 coupé was sent to Zagato's stylists and engineers in Italy. Despite some arguments, the final design was unmistakable on the road, an eye-catching and ear-bending rarity with Zagato's "double-bubble" roof trademark.

Under the bulge of its bulbous bonnet were the four twin-choke Weber carburetors. The 430 rousing horsepower propelled the alloy-bodied coupé to 186 mph (300 kph).

Shorter and lower than the Aston-bodied car, the Zagato was unashamedly only a two-seater – with a padded shelf instead of even token rear seats – but plenty of hand-stitched, top-quality leather maintained the sense of luxury.

The throaty V8 snarl from the cabin, audibly reminded drivers of the car's dramatic performance. Agile, with terrific road-holding, this car was rarely seen on the road.

Right **Aston instrument panels have traditionally imitated the shape of the grille, – this is evident in the Aston Martin Zagato's leather cowl which was used to shade the polished veneer dash.**

Above **Zagato's angular re-interpretation of the famous Aston Martin grille was controversial, but the car was pure adrenalin on the road – two seats, 430 bhp, and the unmistakable howl of a hot V8.**

Buyers had snapped up all 50 limited edition vehicles before the first car was built, knowing that values would soar almost immediately beyond the initial cost of 156,600USD.

Aston Martin insisted that there would only ever be 50 of them, which kept the value high, but a year later they brought out a convertible version. It was arguably more attractive, certainly more expensive at 171,000USD, and again was limited to 50 examples. Later these were even more sought after than the original Zagato, fetching a premium well over the "tin-top" version.

Aston Martin's Virage

V8 | 330 – 465 bhp | 174 mph (280 kph) | 1970s

The 5.4-liter Virage could push 350 lb ft of torque through its rear wheels – enough to propel the two-seater coupé up to 155 mph (250 kph). If this wasn't enough, Aston Martin also offered something extra to its keenest customers, for the Virage, this consisted of a conversion package that boosted the power output by 40%. Once this conversion had been carried out, the capacity of the 48-valve engine increased to 6.3-liters.

Gas-flowed cylinder heads, uprated injection settings, and a hotter camshaft combined to extract a massive 465 bhp, while the torque leapt to 460 lb ft. The abundance of muscle pushed the Virage's peak-speed up to an impressive 174 mph (280 kph), while it could also reach a speed of 100 mph (161 kph) in a little over 11 sec.

To slow the Aston Martin Virage, Aston's engineers employed the largest brake disks ever fitted to a production car at that time; at 14 in. (36 cm) in diameter, they were virtually as big as a complete Mini wheel. Surrounding them were 18 in. (46 cm) alloy wheels clad in 285/45 ZR Goodyear Eagle tires. Like most super cars, the Virage relied on double wishbone suspension at the front, but unusually it had a De Dion rear axle located by a Watts linkage. Its excellent road-holding blends with its overall more relaxed character, as compared to its mid-engined counter-parts.

To drive the Virage fast on open roads took skill and accuracy, and negotiating it through traffic needed care, as the soft aluminum bodywork was easily dented, however, an Aston Martin was truly considered the paragon of comfort and exclusivity.

Left **Following a long line of powerful front-engined Grand Touters, Aston Martin's Virage retains traditional interior appointments, while the shape of the famous Aston grille still adorns the nose.**

Auburn 851 Speedster

8-cylinder | 150 bhp | 100 mph+ (161 kph+) | 1930s

If imitation is the sincerest form of flattery, there can be little doubt that Auburn's various Speedsters have been flattered from the day they were made. The spectacular sporty-looking cars became style leaders in the age of style, and right up to the present day they have inspired numerous replica builders both in America and Europe.

The first Speedster, which was sensationally styled by Count Alexis de Sakhnoffsky, appeared in 1928 on

the 8-cylinder, Lycoming-engined 8-115 chassis. In 1932, Auburn offered the first and last 12-cylinder car ever to be sold for less than 1000USD.

Whether it was the affect of the Depression, or that a Lycoming-engined V12 was being offered at a suspiciously good price, the car didn't sell well, and was dropped in 1934 to make way in for the superior classic Type 851 which was launched in 1935.

The 851, styled almost as a parody of the earlier models by Gordon Buehrig, had the archetypal pointed tail, a huge hood, vast, flexible outside exhaust pipes, and a short two-seater cabin behind a tiny, steeply raked windscreen.

On the dashboard of every 851 was a small plaque, which guaranteed that the car had exceeded 100 mph (161 kph) on being tested by Auburn racer and record breaker, Ab Jenkins.

The car was indeed a staggering performer, with an easily selected option of high or low axle ratios for top speed or acceleration. However, despite this, every 851 was sold at a loss. A very slightly revised model, the 852, replaced the 851 in 1936, but time was running out for Auburn and the 500 or so 851/852 Speedsters failed to revive sales, and eventually lead the company out of business in 1937.

Left **Flamboyant styling makes the Auburn Speedster really stand out from the crowd, while the supercharged, straight-eight engine makes sure it goes as well as it looks. Every car guaranteed to top 100 mph (161 kph).**

Audi 2005T

5-cylinder | 182 bhp | 143 mph (230 kph) | 1980s

For many years, Audi made low-priced, sensible cars, aimed at families that aspired to own a BMW, but couldn't afford one. These cars were well-made in the German manner, which is to say that they lacked in character, but compensated because they were well put together and finished.

Despite their original target driving market, Audi's cars gradually became bigger, faster, and more sought after by an increasingly prosperous middle class in Europe.

For the fast car buyer, who, at the time required a full five-seater with all the basic advantages of an Audi, the Audi 2005T model may have been the

ideal choice. The larger 200 body style gave sumptuous accommodation for five large adults, allowing for every possible comfort, while the drivers had near-sports car performance.

The 2005T could accelerate from 0 to 60 mph (0 to 96 kph) in 8.2 sec, had a top speed of 143 mph (230 kph), and offered overall fuel consumption of just over 19 mpg. This beautifully finished car had only a few causes for criticism.

Below _Vorsprung durch Technik_ ("ahead through technology") is the Audi slogan. The 2005T was an excellent example of the company's advanced engineering and stylish packaging.

Audi Quattro Sport

5-cylinder | 300 bhp | 155 mph (250 kph) | 1980s

Despite its remarkable similarity to the more mundane Audi GT coupé, when it was first launched, the Audi Sport's capabilities removed it from the realm of the merely fast car, and launched it into the class of super-fast, super-expensive car.

The car sat on a wheelbase that is 2 in. (2.5 cm) shorter than that of an Austin Rover Metro, giving it a short, squat appearance which was further emphasized by a longer nose to accommodate the intercooler for the turbo unit.

The engine extracted usable power by means of the exhaust-driven blower. In 1980, after the introduction of this particular engine, Audi upped its power to 130 bhp by using fuel injections. Use of the turbocharger increased this to 200 bhp, with rally cars having enjoyed benefits of up to 350 bhp.

The engine installation in the Quattro Sport represents the fourth generation of Audi turbo refinement, with twin overhead camshafts operating four valves per cylinder.

One very distinct difference from other Audi models was that the Sport had the aerodynamic qualities of a house brick, yet sat

Right **The stubby and purposeful-looking Audi Quattro Sport takes the four-wheel drive format of the original Quattro a stage further, to create a short-wheelbase derivative designed primarily (though not entirely successfully) for rallying.**

on its 225/50VR 15 Pirelli P7 tires, the Sport also had all the charisma of the Ferrari GTO.

The brakes consisted of ventilated disks, and had selectable ABS anti-lock capability. Improved steering also meant that the Sport was easier to drive fast.

The interior promised to be able to accommodate four people, but in practice the Sport was strictly a two-seater.

Austin Healey 3000 MKIII

6-cylinder | 148 bhp | 123 mph (198 kph) | 1950s – 60s

They were affectionately called the "Big Healeys," by almost everyone, because when the Austin Healey 3000s arrived on the scene, they were notoriously larger-than-life.

The family line started in 1952 when specialist car builder Donald Healey exhibited his open, two-seater Healey 100 at the London Motor Show. Powered by a 2.6-liter, 4-cylinder, BMC quickly approved it.

Below **True Brit – the ultimate "Big Healey" – the MKIII was good for around 125 mph (200 kph). 3000s notched up many rally successes. High-speed cruising was no problem, but fast hard driving required plenty of muscle.**

BMC management soon came to an agreement with Healey, and his car went into production in 1953 as the Austin Healey 100. The 4-cylinder cars, including the tuned 100S were built until 1956, when a 102 bhp 6-cylinder 2.6-liter engine was introduced. In a slightly longer, slightly heavier, but still very similar looking shell, this became known as the 100-6.

In 1959, the car underwent a fairly major revision, with some chassis strengthening, front disk brakes, and a capacity increase to 2.9-liters, to emerge as the Austin Healey 3000. With power from a triple-carburetor engine, the 3000 gave an even more vivid performance.

In 1961, an MKII version was introduced with 132 bhp and in 1964 the definitive MKIII appeared, with only two carbs, and quite brutal acceleration in the lower gears.

However, the Big Healey did have some shortcomings. Though capable of being docile, and with its overdrive gearing quite relaxed at speed, the brakes were heavy, it was a brute to drive hard, was short on driver space, and shorter still on luggage space. Moreover, the seats belied their comfortable looks.

Regardless of any complaints, the 3000 built an exceptional reputation as a rally car, and lasted until the end of 1967.

Above **The engine of this Austin Healey 3000 sports a triple-SU carburetor set-up. In this form the straight-six engine puts out 132 bhp.**

Bentley Continental

6-cylinder | N/A | 113 – 119 mph (182 – 192 kph) | 1950s – 60s

The first of the Bentley Continentals was made available in 1952, which set a trend that in 1955 was followed by the S-Type Continental.

Compared with a standard-steel Bentley, the Continental had special coachwork, with four/five-seat accommodation, and usually with two doors. Most of the cars were closed coupés, but an increasing number were ordered with convertible coachwork. Although

many Bentley saloons were still chauffeur-driven, the Continental was essentially a driver's car.

Customers were free to direct the rolling chassis to the coachbuilder of their choice, however, more often than not, H. J. Mulliner and Park Ward, who were both Rolls-Royce subsidaries, along with James Young, were employed.

The S-Type had a new chassis, a longer wheelbase, yet no more passenger accommodation. For the first four years the cars were powered by the long-established 4.9-liter straight 6-cylinder engine, but from late 1959, a new light-alloy V8 engine took over.

Except for a very few 1955/1956 models, all models had the Rolls-Royce/GM automatic transmission. Every car had drum brakes, boosted by the famous Rolls-Royce mechanical servo.

The S-Types were all large, heavy, and impressive machines, with power-operated soft-tops, wind-down windows in the doors, and the highest possible standards of trim, decoration, and seating.

Left **The Bentley Continental S1 was built between 1956 and 1959, and was the last of the Continentals to use a 6-cylinder engine (in line with 8:1 compression ratio); a V8 was used for the S2 and S3 models.**

Bentley Mulsanne

V8 | N/A | 135 mph (217 kph) | 1980s

It took Rolls-Royce 51 years to introduce a Bentley that was more than a re-badged, re-radiatored Rolls-Royce. The Bentley that was eventually introduced was the Mulsanne, named after the famous straight on the Le Mans racing circuit – the scene of the beginning of the old Bentley legend.

The Mulsanne was, at last, a really fast Bentley, one that W. O. Bentley himself would certainly have been proud of. W.O. always disapproved of supercharged engines, feeling that if

more power was required, he could easily provide the engine with more capacity, while retaining all the flexibility and reliability that his cars were renowned for.

The Mulsanne changed all that. Its 6.75-liter V8 engine produced an undisclosed amount of power which was sufficient enough to propel the heavy car to speeds of over 130 mph (209 kph), and from 0 to 60 mph (0 to 97) in 7.4 sec.

This car model sat four people in great comfort, and could transport them with a smoothness and speed that was then considered nothing short of uncanny. Handling was considered "soft," with rather too much roll in cornering. However a Bently was not usually driven like a Ferrari, and with normal Bentley-style driving, was a delight to handle, – it was recognized during its rein, as a very expensive delight.

Left **Ettore Bugatti once described the Bentleys which beat his own cars at Le Mans in the 1920s as "*les plus vites camions du monde*" – the fastest cars in the world. No lorry, however, was ever as fast or as luxurious as the Bentley Mulsanne, named for the famous straight on the Le Mans circuit where Bentleys had their finest hour.**

Blower Bentleys

4-cylinder, sohc | 250 bhp | 130 mph (209 kph) | 1930s

If there is any single universal image of the vintage sports car, the "Blower" Bentley undoubtedly has it.

In 1930, in an effort to keep his Bentleys ahead of the opposition, Sir Henry Birkin, the marquee's most dashing champion, commissioned engineer Amherst Villiers to supercharge the 4.5-liter engine.

W. O. Bentley disliked the idea, but Birkin and the racing "Blowers" brought a final rush of fame to the winged "B," pitting the massive British touring cars against the best of Continental racing machinery.

There were only four of the real, racing "Blower" Bentleys, each of them different. The first became a single-seater, the second a short-wheelbase four-seater, the third a longer four-seater, and the last, completed in 1930, a short-chassis car, which finished second in the 1930 Pau Grand Prix, driven by Birkin himself.

They are not easy cars to drive. One has to admire the stamina of Birkin and the others who wrestled with the heavy steering, stubborn crash gearbox, and jarring ride for long hours in endurance races.

The road holding was surprisingly good, but there was no disguising how much kinetic energy the huge drum brakes had to destroy from very high speeds.

Apart from Birkin's team cars, a limited run of 50 Blowers was built, and their rarity and performance make them especially valuable today.

Left **The classic supercharged 4.5- liter Bentley.**

Below **Despite W. O. Bentley's opposition to the idea, and their relative lack of racing success, the Blower 4.5 has become, arguably, the most glamorous Bentley.**

BMW 3-Series Cabriolet

4 or 6-cylinder | 90 – 170 bhp | 100 – 130 mph (161 – 209 kph) | 1970s

In 1966, BMW introduced its first small/medium two-door sedan car, the unit-construction 1600-2. Like all such modern BMWs, this had independent suspension front and rear, and was immediately recognized by the characteristic kidney-shaped radiator grille.

Eventually this car was sold with a whole variety of more powerful engines, and was not supplanted by a new model, the first of the famous 3-Series cars, until 1975. In 1982, a reworked, but essentially similar "Mark 2" 3-Series model came along and sold even better than the original.

In the 1970s, BMW was only equipped to build conventional sedan cars. Demands for an open-top car therefore could only be satisfied by cooperating with the Stuttgart-based company of Baur. It was not until 1977 that Baur unveiled its first 3-Series Cabriolets, which stayed in the market until the mid-1980s.

Right **The 3-Series Cabriolet might be an open-top car, but it has all the equipment of the Munich sedan, including sumptuously trimmed seats.**

Above **BMW's 325i Cabriolet, which went on sale in 1986, was the first "in house" design from BMW for many years. The previous soft-top 3-Series car was built by Baur, and had permanently fixed body sides and a permanent roll-bar over the seats.**

Baur's Cabriolet was essentially a conversion of the 3-Series sedan. Not only to give roll-over protection, but also to retain body shell rigidity, the Baur-designed Cabriolet had rigid cant-rails over the top of the doors, permanently fixed door pillars, and a pressed "roll-bar" over the passengers' heads. There was less space in the rear seats than in equivalent sedans.

BMW, however, was not completely satisfied with Baur's solution to the conversion, and so the company eventually designed its own car. The car's floorpan was significantly stiffened, and with the soft top furled, there was complete all-round visibility. The factory-designed Cabriolet went on sale in 1986, and completely replaced the Baur unit.

BMW 328

6-cylinder | 80 bhp | 95 mph (153 kph) | 1930s – 40s

Considering the strength of BMW's high-tech sporting image today, it is difficult to believe that the company's first venture as a car manufacturer was in building a version of the humble Austin Seven.

BMW was already well known for its motorcycles and aero engines before it acquired the German manufacturing rights for the Seven. The acquisition was made when BMW took over Dixi in 1928 and relaunched Dixi's interpretation of the Seven as the BMW 3/15, on New Year's Day 1929.

Following this intial venture into the car manufacturing world, the company's real sporting fame sprang from the 1933 arrival of its first 6-cylinder model, the 303 sedan – a

tiny car for a 6-cylinder, with just 72 ci (1175cc) and about 30 bhp to its name.

This car soon evolved into the 315, and then the 319, which was the final forerunner of the classic 328. Unveiled in the summer of 1936, the 328 was based on the 319's well-proven tubular chassis, with an engine of close to 2-liters and a clever hemispherical head design, with inclined valves operated by a single camshaft, and a crossover pushrod arrangement. This gave the new engine most of the advantages of a real twin-cam design, but without the complexity and manufacturing expense.

The engine quickly earned a reputation for being both reliable and crisp. Additionally, it set new standards of road holding and handling – with much of the refinement of a sedan car ride combined with standards of grip found in racing cars. This is largely due to the 328's brave insistence on using independent front suspension at a time when beam axles were generally the order of the day on sports cars, but it said even more about the overall integrity of the chassis design.

To complete this magnificent mechanical package, BMW engineers also came up with an aerodynamically efficient body, which made most other sports cars of the day look positively primitive. It even had a smooth full-length under-tray to dramatically cut drag below the car.

Left **The BMW 328 was considered both elegant and superbly engineered. Its advanced, wind-cheating bodywork looked good, while independent front suspension gave the 328 the comfort of a sedan car ride with racing car levels of roadholding.**

BMW 745i

6-cylinder | 295 bhp | 150 mph (241 kph) | 1980s

The 745i was a 3.5-liter turbocharged 7-Series car. Luxuriously appointed, it had an onboard computer, ABS braking, and very high levels of trim and equipments for its time.

The engine's performance was near sensational; completely free of the dreaded turbo-lag. The ABS braking system handled slippery conditions with ease and safety, and the steering further emphasized the overall excellent nature of the car. Its special attraction lay in its lower price and less prohibitive insurance cost, which for many people balanced out its relative lack of go.

Below **The flagship of the BMW fleet was the large, luxurious 7-Series sedan. The turbocharged 745i topped the range in Germany and offerd 150 mph (241 kph) transport in the grandest manner.**

Alpina BMW

6-cylinder | 185 bhp | 133 mph (214 kph) | 1980s

***Above* Among the many companies which tune BMWs for road and track use, Alpina was probably the best known, and in racing terms the most successful. Their Cl and C2, (which eventually replaced the Cl), were both based on the compact 3-Series.**

The rapid Alpina C cars were based on the 3-series BMW. The first was the C1, based on the BMW323i but with a much modified version of the 141 ci (2,316cc) 6-cylinder engine. An increase in power output from 150 to 170 bhp gave substantially improved performance, matched by very effective suspension changes.

The C2, the successor to the C1, was even quicker, attaining a top speed of 133 mph (214 kph) with the five-speed Getrag sports gearbox and accelerating from 0 to 60 mph in 6.6 sec. This level of performance was achieved by using a 152 ci (2,490cc) engine derived from the BMW525 Eta, with special Alpina cylinder head and crankshaft, and Bosch LE Jetronic fuel injection.

To cope with the extra surge, Alpina fit progressive rate suspension with Bilstein gas-filled shock absorbers, a limited-slip differential, and 195/50 VR16 Pirelli tires.

BMW M1 Coupé

6-cylinder | 277 bhp | 160 mph (258 kph) | 1970s

In 1972, BMW unveiled a remarkable, gull-winged, mid-engined coupé BMW Turbo on the show circuit. Although never intended for production, in 1978 a car with most of the Turbo's style and character was unveiled as a production reality.

It was called, simply, the M1 – the M-Style logo being BMW's way of indicating "Motorsport." The car, powered by BMW's potent 24-valve, 3.5-liter twin-cam straight-6 motor, was less radical and more practical than the show-going Turbo coupé. It had conventional doors and striking styling by the ubiquitous Giugiaro. It was openly admitted that it was intended first as a racing car, but the production run also included road cars, fully trimmed and well equipped.

The majority of the 450 or so M1's built between 1978 and 1980 made outstanding road cars. Where racing engines could boast 500 bhp (as much as 800 bhp with turbocharging), the fuel injected production engine gave a perfectly adequate 277 bhp and comfortable 160 mph (257 kph) performance. Even on softer springing and road tires, the M1 retained much of its racing poise. If bureaucracy had rendered the M1 something of a failure in its intended racing guise, nothing has taken away its brilliance as a road car.

Below and Right **The M1 was developed to help BMW enter into racing, but changes in racing regulations later made it obsolete.**

BMW M535i

6-cylinder | 218 bhp | 143 mph (230 kph) | 1980s

In order to meet the challenge of the new 16-valve Mercedes-Benz 190E 2.3 in the smaller car sector, BMW introduced their M535i.

The BMW M535i had a 209 ci (3,430 cc) single-overhead-cam fuel-injected straight-6 engine. Acceleration from 0 to 62 mph (0 to 100 kph) was possible in 7 sec.

There was a choice of three different gearboxes for the M535i – the BMW 5-speed with overdrive unit, a close-ratio Getrag 5-speed, or the ZF triple-range switchable automatic.

The braking for this car was accomodated by four-wheel ventilated disk units, with ABS anti-lock control. Rear seat accommodation was much more generous, with proper leg room for two adults.

Below **M-style engineering and cosmetic packaging turned BMW's originally rather staid 5-Series sedan into the road-burning 143 mph (230kph) M535i. In this guise, the comfortable five-seater was thought at the time to seriously embarass many other sportscars.**

BMW M635CSi

6-cylinder | 286 bhp | 158 mph (254 kph) | 1980s

The BMW M635CSi coupé was introduced in 1985. The in-line 211 ci (3,453 cc) engine has duplex-chain-driven double overhead cams, four valves per cylinder, and a compression ratio raised to 9.6:1.

Digital engine electronics controlling the ignition and fuel supply systems had been programmed to give optimum timing and fuel metering under all conditions. An ABS, anti-lock braking system was fitted as standard.

One particularly interesting feature of all the big BMWs was the availability of the ZF 4-speed automatic gearbox, with its three ranges of operation.

***Above* BMW Motorsport division's engine and suspension modifications turned what was considered a rapid 635CSi, into the incredible M635CSi.**

Bristol Beaufighter

V8 | N/A | 150 mph (241 kph) | 1980s

The three cars from the Bristol Car Company – the Britannia, the Beaufighter, and the Brigand – were named after famous Bristol aircraft from the company's earlier days.

The most glamorous of the Bristol range was the turbocharged Beaufighter, which shared the same version of the Chrysler hemi as the Brigand sedan. The Beaufighter, however, was a much more distinctive and rather angular car, distinguished by its flamboyant targa-type top, with a substantial roll-hoop and a lift-off center section.

The car was styled by Zagato and introduced as the 412 Convertible, later known as the Convertible Saloon, in 1975. Also among its styling features was the traditional Bristol way of housing the spare wheel – in a concealed compartment in the front wing. The turbocharged version and the Beaufighter name were launched in 1980. Like the similarly powered Brigand, it reached 0 to 60 mph (0 to 97 kph) in 5.9 sec.

Top and Left **For such a large, heavy, and well-equipped machine, the Bristol Beaufighter was surprisingly light in feel and easy to drive.**

Bristol Brigand

6-cylinder | 277 bhp | 160 mph (250 kph) | 1970s

One of the most expensive of the Bristol range of cars is the Brigand, a turbocharged two-door four-seater sedan whose sleek lines date back to 1978 and the introduction of a car that was then known as the 603 sedan.

It has the ability to reach 60 mph (97 kph) from rest in less than 6 sec. The Brigand may not be as quick as

the Vantage, but will comfortably outrun the "ordinary" Mulsanne.

Part of the Bristol's strength lies in its chassis performance, where the engineering quality is evident. When it was first launched into the modern super car world, its ride and handling were considered quite exceptional for such a luxurious and spacious sedan.

Below **Although many other Anglo-American hybrids have been hastily conceived and appropriately short-lived, the 140 mph (225 kph) Bristol Britannia sedan showed that a marriage of the best British engineering and an American mass-produced power unit could result in a genuine classic car.**

Bugatti EB110

V 12 | 550 bhp | 217 mph (349 kph) | N/A

They say money isn't everything, and if you wanted to be part of the renaissance of the Bugatti marquee, you would have soon found that this was true. Despite the mystique that built up around his beautiful racing and sports cars before World War II, the company effectively died with its founder Ettore Bugatti in 1947.

Forty years later, a business group bought the rights to the Bugatti name and set to work on one of the most intense and talked-about new car projects ever seen to that date. From a standing start they set to work on a super car worthy of carrying the oval badge with the EB initials – the EB 110.

In a striking new factory outside Modena, the team started by designing a novel engine. This engine, a V12, had four turbochargers that

facilitated instantaneous response. Aided by the turbochargers, the 3.5-liter, 60-valve unit spits out 550 bhp and can rev to 9200 rpm. Within the carbon-fiber composite chassis, the inline block was offset to the left with the 6-speed gearbox alongside, driving all four wheels. Computer-controlled, semi-active suspension pressed the Bugatti down to the road as the speed climbed.

The EB110 was a low, compact two-seater in the Lamborghini mold, rather than a front-engined fast tourer of Bugatti legend.

Left and Above **A glorious name reborn; the EB110 Bugatti outside the high-tech factory in Modena where it was built.**

Bugatti Atlantic

Twin-cam, supercharger | N/A | 130 mph (209 kph) | 1930s

Europe's most flamboyant coachbuilders loved to dress Bugatti chassis, and the Type 57 received some outrageous as well as some very beautiful car bodies. Their appeal varies, though all are valuable, but if forced to pick one model of Type 57, many connoisseurs would select an SC with Ettore Bugatti's son's amazing Atlantic coupé body.

The Atlantic is streamlined to feature unique external riveted

flanges on the roof and wings. This style also figures in Ralph Lauren's impressive stable. The Atlantic is a dream to drive, but only in cool weather – the reason being that it has no ventilation, and the heat from the underfloor exhaust makes it stifling hot.

In fact, only three Atlantics were built, and though they have only changed hands privately, rumors suggest that one of the most recent buyers paid more than seven million dollars to acquire it.

Below **Ettore Bugatti's son Jean, later tragically killed driving one of his father's cars in 1939, displayed his inventive design skills with the amazing Atlantic coupé on the 57SC chassis.**

Bugatti Royale

8-cylinder | N/A | N/A | 1930s

The Bugatti Royale is one of the rarer, more fabled, and more valuable of the mark's cars. It was Bugatti's "Car of Kings," and was intended to appeal to the crowned heads of Europe.

In fact, Bugatti only ever sold three Royales, while the family retained another three for their own use. It was the most expensive car ever sold. Fifty-six years later, one of the six, with a body by French coachbuilder Kellner, went for auction at Christies in London. The hammer fell on a bid of 9.9USD (5.5GBP) million, an all-time record for an automobile which stood for three years.

All six Royales still exist, each with a unique body style. It is an enormous car; the 8-cylinder, single-cam engine displaces no less than 12.7-liters, and the cast aluminum wheels are capable of carrying a truck.

Yet only one of the cars produced looked like a limousine. Its owner, a French industrialist, furthermore requested that Jean Bugatti design it without headlamps, as he did not intend to drive it after dark.

Bugatti had hoped that the Type 41 Royale would secure him the title of the builder of the greatest car in the world, therefore eclipsing Rolls-Royce, Hispano-Suiza, and Isotta-Fraschini. Commercially it was a memorable failure – the reason for its faliure being that it arrived on the market at a time of depression. Historically, however, it guaranteed Ettore Bugatti exactly the reputation he always craved – that of the creator of one of the biggest and costliest cars in the world.

Left **The car built for kings – even though their majesties declined to buy. Ettore Bugatti retained this Royale for his own use. Today it is one of the stars of the astonishing Schlumpf museum in France.**

Above **The massive Bugatti Royale Type 41 was not a sale success. It failed to find favor with royalty and heads of state for whom it was intended.**

Bugatti Type 43

8-cylinder | 120 bhp | 110 mph (177 kph) | 1920s – 30s

The Type 43, introduced in 1927, used a very slightly detuned version of the Type 35B Grand Prix car's supercharged 2.3-liter straight-eight engine in a chassis derived from the underpowered Type 38 tourer. The main difference between the cars was the Type 43's shorter wheelbase.

What this created was the world's first 100 mph (161 kph) production car – which was significant at a time when 70 mph (113 kph) was considered quick for production cars.

It was sold, very expensively, as a 3.5-seater sports tourer. The engineering was typically Bugatti; the engine, square-cut and plain looking, concealed its artistry inside. It also concealed one of Bugatti's weaknesses, a built-up crankshaft with roller bearings and no forced lubrication. This crank was appealing to look at but, like much of Bugatti's

Right **This was the famous late 1930s SS100, complete with twin spare wheels, a slab tank and removable side curtains.**

work, needed constant attention to keep it serviceable.

Through the slick, 4-speed gearbox and the supercharger taking effect above about 1500 rpm, it would reach 60 mph (97 kph) in less than 12 sec, and had a top speed of 110 mph (177 kph). This performance was matched by a superb chassis.

Right **The Bugatti Type 43 is quite literally a racing car for the road. The first production car to be capable of more than 100 mph (161 kph), its supercharged 8-cylinder engine was capable of whisking it to a 110 mph (176 kph) maximum in fourth gear.**

Bugatti's Type 57

Straight 8, twin-cam engine | N/A | 130 mph (209 kph) | 1930s

First seen in 1933, the Type 57 had a 3257 cc (199 ci) straight-eight, twin-cam engine. Originally the 57 could not quite reach 100 mph (161 kph), so the factory offered a supercharger and called it the 57C (Compressor). Despite its conventional leaf-spring layout, the chassis proved to have fine handling, which was further improved in the 1936 sports version, the 57S.

But the *crème-de-la-crème* of the range is a sports chassis with the supercharged engine – the 57SC, capable of over 130 mph (209 kph).

A mere 30 were initially built, ranking the Bugatti Type 57 rank with the Alfa 2900B in both performance and rarity.

Below **Every Bugatti excites collectors, but Type 57 along with the 57SC (sports chassis with supercharger) are the most desirable Bugatti cars to collect. The example shown here has a coupé body by London coachbuilder Corsica.**

Cadillac Allanté

V8 | 170 bhp | 124 mph (200 kph) | 1980s

From a mechanical point of view, the Allanté had a big, well-developed V8 engine, in a transversely positioned package, which was driven through a 4-speed automatic transmission to the front wheels. Front and rear suspensions were independent.

Despite the wheelbase being 99.4 in. (252.5 cm), Cadillac provided only two seats to this car, with a great deal of useful stowage space behind the leather-trimmed seats. Almost everything was electrically operated – including soft-top, windows, and seat positioning – and the car was considered so well equipped for its time that the most significant equipment option was a cellular phone.

Pininfarina produced a classic body for the car, which was clearly right for the sector it was aimed at. The drag coefficient, 0.34, was considered excellent for an open-top car at the time, and was made possible by the detailing of headlamp positions, spoilers, and massive energy-absorbing bumpers. The interior was mostly in carefully crafted leather, while the fascia seemed to have more switches and controls than the average jet aircraft.

For those who could not stand breeze in their hair, Cadillac also offered a lift-off hardtop for the Allanté, which made the car look even more sleek than when it was open, and further reduced drag and fuel consumption.

Left and Below Ten years on from the Eldorado, Cadillac produced the smaller, more elegant, and Pininfarina-styled Allanté. Though still a fast car, the Allanté was more practical than the obsolete model.

Cadillac Eldorado

V8 | 190 – 235 bhp | 115 – 125 mph (185 – 201 kph) | 1970s

In 1966, based on the general layout of the Oldsmobile Toronado, Cadillac introduced its first ever front-wheel-drive car, the Eldorado Coupé. Compared with any previous Cadillac, the front-drive Eldorado was astonishingly different, advanced, and technically courageous.

For the first few years the Eldorado was sold only as a closed coupé, but a convertible version was offered in 1971. In those years it was Cadillac's only open-top car.

GM's solution to the front-drive packaging problem they encountered was to sit the big V8 engine above the line of the front wheels, to mount automatic transmission and final drive components alongside it, and to take the drive from engine to transmission through a massive and carefully developed chain.

The car was extremely long, with large overhangs at front and rear. The hood had a upright and assertive prow, with a pair of headlamps tucked at each side of the rectangular grille. The hood and the boot lid were both long, and the line of the car swept smoothly from nose to tail.

As the years passed, the style of the Cadillac Eldorado was only slightly refined, with "opera" windows installed in the hardtop version. By contrast, the grille style seemed to change every year.

Right **The 1973 Eldorado convertible.**

Right **The 1971 Eldorado convertible. Body styling on the Eldorado was sculptured and featured a rise over the rear fenders.**

Chevrolet Corvette

6-cylinder | 150 – 360 bhp | 150 mph (241 kph) | 1950s – 60s

The Chevrolet Corvette was the first two-seater sports car designed by General Motors. From 1953 to 1962, all Corvettes were built on the same basic 102 in. wheelbase chassis, with unfortunately floppy suspension. 6-cylinder engines were only used between 1953 and 1955. Chevrolet's new lightweight "small-block" V8 engine was introduced in 1955, and was fitted to all cars from 1956 onward. Although original Corvettes were only built with automatic transmission, a manual option followed in 1956.

The original glass-fiber style, used until 1955, had two headlamps recessed in the front wings, a "mouth-organ" grille, a wrap-around windscreen, and detachable sidescreens. For the 1956 version, there was a completely new, larger, and more stylish body shell, with heavily sculpted sides behind the front arches, and wind-up windows; for the first time there was the option of a detachable hardtop with wrapped-round rear window. Fuel injection was offered in 1957.

The same basic body, modified to be 3 in. wider than before, and with four headlamps, was launched in 1958. This style was maintained, with annual facelifts, until the end of the 1962 selling season.

A Corvette with high powered V8 engine (which was optional), some of which were further equipped with fuel injection and aluminum cylinder heads, was a fast car by any standard of its time. To a whole generation of young Americans, this type of Corvette was the definitive open-top American car of the period.

Left **Corvette – this was the instrument panel of the 1962 model.**

Right **Corvette – the all-American sports car with fiberglass bodyshell and Chevy V8 engine.**

Citroën CX25 GTi Turbo

4-cylinder | 168 bhp | 136 mph (219 kph) | 1980s

The French have a history of designing and manufacturing cars that is just as extensive as any other. Two of their fastest machines during the 1980s included the Citroën GTi Turbo, a normal production car, and the other a homologation special, the Peugeot 205 Turbo 16.

Both cars were genuine over-125 mph (201 kph) cars. Citroën claimed that their CX GTi Turbo could reach 136 mph (219 kph). Furthermore, the CX's turbo installation, then reputed to be the very best available, gave the venerable 2.5-liter 4-cylinder engine a whole new lease of life.

A new rear spoiler served to maintain a decent drag figure of 0.36 which, in conjunction with the increased urge of the engine, up

from 138bhp to 168bhp with the Garrett turbo and Bosch fuel injection, gave the car its high-speed image. Acceleration was also good for such a big car, at only 8.6 sec for 0 to 60 mph (0 to 97 kph).

The car had all the usual CX advantages, including a comfortable interior, the long-legged ability to eat up the freeway miles, the excellent, but just a little too sensitive, all-round disk braking system, which made light of stopping the 3,053lb car from any speed.

As with previous Citroën models, the CX25 GTi-Turbo was not really a car to jump into without thinking – this model rewarded its driver for taking the time to acclimatize to the car's specifics, especially the instrumentation and controls. However, once more extensive knowledge of how to use the car was acquired, the Citroën proved to be a very rewarding car to own and drive.

Below **It may not have been the most obvious of sporting cars, but the front-wheel-drive Citroën CX25 GTi Turbo showed that the makers of the 2CV had not forgotten the opposite end of the market. The Turbo's top speed was almost exactly double that of the much loved *deux-chevaux*.**

Citroën DS Cabriolet

4-cylinder | 78 – 106 bhp | 93 – 107 mph (150 – 172 kph) | 1960s

Citroën's DS19 was a machine with exciting specifications – not only did it have front-wheel-drive and a wind-cheating styling which was far ahead of its time, but it also had independent self-leveling suspension by high-pressure hydro-pneumatic units, four-wheel disk brakes, and full-power steering with power generated by a pump on the engine. From 1965, more powerful units for the DS21, as well as all of Citroën's subsequent models, were introduced.

Citroën itself concentrated on the building of four-door sedans and lengthy five-door station wagons. In 1959, however, the French coachbuilder, Henri Chapron, showed a convertible version of the DS19 at the Paris Salon, and a year later this model was officially adopted by Citroën.

Chapron's *décapotable* (convertible) was a thorough conversion of the

Below **With its Cabriolet top furled, the chapron-style Citroën was a smart open-air machine. The sedan's long wheelbase was retained, and there was still ample room for four occupants.**

Above **Years before the rest of the world discovered aerodynamic styling, Citroën was practicing the art of air-flow management. Launched in 1955, the basking shark style of the Citroën DS was still efficient in the 1970s.**

existing DS sedan. Ahead of the windscreen, and below the waistline, the styling of the convertible was the same as that of the sedan; and was so meticulously done that some people failed to notice that there were two passenger doors instead of four!

The Chapron Cabriolet kept all of the Citroën's idiosyncratic mechanical features, including the one-spoke steering wheel, the strange transmission, and of course, the self-leveling suspension. The same shark-like nose and prominent headlamps at the front of each wing remained, although the doors were much longer than those normally fitted at the front of a Citroën in order to allow access to the rear seats.

The soft-top, when raised, fitted neatly around the wind-up windows, while some cars were conversly fitted with detachable hard-tops.

Over the years, the Cabriolet's equipment changed along with that of the sedan. The most obvious visual change was the 1967 adoption of a lower four-headlamp nose, in which one pair of lamps was connected to the steering, and swiveled when the wheels turned.

Cizeta

V16 | 520 bhp | 190 mph (306 kph) | N/A

The Italian super car Cizeta's name is simply derived from the initials in Italian of the man who conceived and developed it, Claudio Zampolli.

The Cizeta was not quite a 200 mph (322 kph) car but was unique because of one mechanical innovation – a V16 motor. Styled by Marcello Gandini – who also created the Diablo – the Cizeta V16T was a similar mold.

Its aluminum body clothed a tubular steel chassis with an unusual transmission layout. The big engine sat sideways across the car, but the gearbox projected longitudinally behind, driven from the middle of the block. This effectively made the engine two linked V8s, with four twin-cam cylinder heads, making eight camshafts, and an amazing 64 valves! With 6-liter capacity, the Cizeta had electronic ignition and fuel injection, and was fitted with a catalytic converter.

Inside, the Cizeta was considered luxurious, with leather trim and a comprehensive climate-control system.

Below **Low, sleek, and mean – the super-high-tech Cizeta came from Modena, Italy, the home of Bugatti and Ferrari.**

Colt Starion

4-cylinder | 168 bhp | 137 mph (220 kph) | 1980s

The Mitsubishi Colt Starion was a Japanese performance coupé which used a turbocharger to boost its power.

Front and rear spoilers aid aerodynamics and combat lift at speed, but had been carefully designed not to detract from the clean shape of the car. The Starion's steering was of recirculating-ball type. Additionally, a 5-speed manual gearbox was standard, with ratios that complement the blown 4-cylinder engine's power output.

Above **Colt's considerable experience and success in production racing is evidenced by the company's reputation for building rapid and rugged cars for the road, from small turbo hatchbacks to this top-of-the-range Starion coupé.**

Handling was good, being firm and slightly choppy as the all-round MacPherson strut-type suspension was more biased to handling than to ride comfort. Unlike many Japanese cars, the Starion had real character, and was enormous fun to drive.

Cunningham C3 Continental Coupé

V8 | 210 bhp | 138 mph (222 kph) | 1950s

Briggs Swift Cunningham launched the road car concept in 1951 when he planned to build Cunningham C2 production sports cars alongside C1 racers in what were then his new Palm Beach workshops.

In 1951 four cars were built although only one was a road car with a Cadillac V8, the remaining racers had Chrysler Hemis. From February 1952 Cunningham cataloged a Chrysler-powered, Vignale-bodied coupé, the C3 Continental.

Although fast (at almost 140 mph [225 kph]) and expensive for its time, the Cunningham road cars sold just 18 coupés and nine convertibles before production petered out in 1955.

Below **Advertized during the 1950s as "the ultimate in sports cars," the Cunningham C3 Continental Coupé was the roadgoing version of the racing cars built to challenge the likes of cars such as Jaguar at Le Mans.**

Delahaye Type 135 Convertibles

6-cylinder | 130 – 160 bhp | 95 mph (153 kph) | 1930s – 40s

Not only was the Type 135 the fastest Delahaye ever produced, but it also had what was then considered a thoroughly modern chassis. At a time when many large and fast cars still relied on beam axle front suspension, the latest Delahaye used coil spring independent front suspension.

The new car's wheelbase was a generous 116 in. (295 cm). Although some cars retained free-standing headlamps, many had contoured fronts, no running boards, and sharply swept tails.

In the late 1930s, just before WWII, these cars came to be associated with the social decadence and privilege of the time.

Above **French engineering *savoir-faire* linked to elegant British coachbuilt style. Note the serviceable, authentic soft-top irons, and the rear-hinged passenger doors.**

Left **Featured here is the British drop-head coupé body style by Pennock. French styles were often more stark and purposeful.**

De Tomaso Pantera

V8 | 330 bhp | 160 mph (258 kph) | 1960s – 70s

The former Argentine racing driver Allejandro de Tomaso built some reasonably successful cars before he built the Pantera. However, the Pantera soon came to overshadow all of his previous models.

In 1967 de Tomaso bought Ghia, the firm known for styling the Mangusta. When Ford subsequently took over Ghia, Tomaso found an unlikely but enthusiastic ally.

With Ford's help, the mid-engined basics of the Pantera were refined, given an attractive new body, and relaunched at the 1970 New York Motor Show as the De Tomaso Pantera, "Powered by Ford."

The 5.7-liter V8 and 5-speed ZF transaxle were mounted in a subframe in the back of a steel, unit-construction two-seater coupé shell. It stood on coil spring and wishbone suspension, with ventilated disk brakes all round, and larger back tires than front tires. It was a big car at 14 ft (4.3 meters) long.

Below **The De Tomaso Pantera GTS, styled by Ghia and powered by Ford. Good looks and handling, and a midmounted engine made it a car to be reckoned with in its day.**

Duesenberg J and SJ Convertibles

8-cylinder | 200/320/400 bhp | 135 mph (217 kph) | 1920s – 30s

Armed with the engineering skill of Fred Duesenberg and the entrepreneurial flair of E. L. Cord, in 1928, the Duesenberg Company launched its Model J.

Its 6.9-liter, 8-cylinder, twin-camshaft Lycoming engine boasted four valves per cylinder, as well as an output of 265 bhp, which was then streets ahead of its rivals. In addition, it came as a bare chassis which could be clothed to the purchaser's taste.

In 1932 the SJ, a supercharged model, appeared and quickly became the most coveted Duesenberg. Of the fewer than 500 Model J's produced, only 36 were SJ versions.

Although American film stars such as Clark Gable and Gary Cooper took delivery of short-chassis SSJ's (with roadster bodies), the financial collapse of E. L. Cord's empire finally brought this fabulous car to an end.

Above **Born into the age of decadence, it didn't seem to matter that this huge 8-cylinder twin-cam engined car only had two seats – what was important was the character of the car, and the attitude it conveyed about its creators and drivers.**

Ferrari 400i

V12 | 315 bhp | 152 mph (245 kph) | 1980s

As an ex-racer and racing team manager, Enzo Ferrari's thirst for great engine performance initially influenced the design and production of his customer cars.

Yet with science introduced into the closed shop of racing car design some 40 years ago, and Lotus's late founding-genius Colin Chapman, then at the forefront of the field, Ferrari was forced to take into account and maximize on other dynamic aspects of its race cars.

As with so many developments at the infamous Ferrari factory at Maranello, these other aspects

eventually filtered down the production line, and are now considered responsible for keeping road cars in the vanguard of high-performance vehicles.

An example of one such model was the company's high-performance 400i model, a full four-seater two-door car that was powered by a 294 ci (4,823 cc) fuel-injected V12 engine.

The car had the usual classic ageless Pininfarina styling, and despite having been in production for many years, still looked as up-to-date as any other performance car of its era. The car has a top speed of 152 mph (244 kph). It is still considered one of the great classic cars of its time.

Below **The supremely elegant 400i offers 150 mph (241 kph), plus Ferrari motoring for four, and was the first Ferrari ever to offer automatic transmission.**

Ferrari GTO

V8 | 400 bhp | 190 mph (306 kph) | 1980s

The GTO, which stands for Gran Turismo Omologato, was powered by a 174 ci (2,855 cc) 32-valve V8 engine with twin turbochargers and could develop a massive 400 bhp at a typical Ferrari engine speed of 7,000 rpm.

The turbo system and its ancillaries used much of the technology developed on Ferrari's turbocharged Grand Prix cars. The transmission was entirely designed and made by Ferrari, and the single composite unit comprised the five-speed gearbox, clutch, and limited-slip differential.

Four-wheel ventilated disk brakes were fitted, but ABS was not deemed necessary at the time. Suspension

followed the Grand Prix car layout, and was independent all round, with coil springs and Koni dampers providing the springing medium. Slightly different wheel sizes were employed at front and rear, with 16 x 8 in. (41 x 20 cm) front wheels and 16 x 10 in. (41 x 25 cm) wheel size at the rear. Goodyear NCT tires were supplied as standard fittings.

The body styling was very similar, but not identical, to the 308GTB Ferrari, and extensive use was made of wind tunnel testing to achieve a very aerodynamic shape.

On the performance front, the GTOs acceleration was 0 to 62 mph (0 to 160 kph) in less than 5 sec., with a top speed of 190 mph (306 kph).

Below **The Ferrari GTO, introduced in mid-1984, amply underlined Ferrari's commitment to ultimate performance – having finally overtaken the legendary front-engined V12 Daytona as what was then the fastest Ferrari road car.**

Ferrari Testarossa

12-cylinder | 390 bhp | 181 mph (291 kph) | 1980s

During the 1980s, the Ferrari Testarossa was considered one of the most beautiful competition cars ever to have raced. The car's unique name was acquired when its cam covers were painted red instead of the more usual black.

In the year of the original car's introduction, 1958, it won the 1,000 km of Buenos Aires, the Sebring 12 hours, the Targa Florio, and Le Mans. During the next year, success followed success for this machine, restoring its 1985 reputation as a high-performance Ferrari car.

The car's power came from a 302 ci (4,942 cc) flat-12, 4-valve-per-cylinder engine mounted amidships in the car, and on top of the transmission. This impressive powerhouse produced 390 bhp at 6,300 rpm, to give the 1985 Testarossa acceleration from 0 to 62 mph (0 to 100 km) in 5.8 sec.

The stunning body was from the studios of Pininfarina. The new radiator position, in the middle of the

***Right* Under those sweeping lines lies a flat-12 power unit which can push the Ferrari Testarossa to 194 mph (312 kph) with stability and safety. And to match the performance in speed, the huge 18 in. (46 cm) wheels contain massive "race-bred" braking power.**

Above **The Ferrari Testarossa lived up to the sporting pedigree of its predecessors. The wind-tunnel-developed body made the car extraordinarily stable at high speeds.**

Below **Restrained comfort in the cockpit of the Ferrari 512TR; high-quality leather everywhere, and the famous ball gear-knob and exposed metal gate located between the seats have typified Ferraris since the 1950s.**

car instead of in the more usual place at the front, had been accommodated successfully in the body styling. The car was also very aerodynamically stable at all speeds, an ability that had come to be expected of all Ferrari GT models at the time.

The Testarossa's interior marked a dramatic improvement in driver and passenger accommodation, but most notable about this car were the ergonomics that gave it speed.

Ferrari 308GTB QV

V8 | 240 bhp | 158 mph (254 kph) | 1980s

During the 1970s, Ferrari decided to replace its successful V6-engined Dino series. However, they wanted to retain the same basic mid-engined chassis layout, but produce a new engine. Starting in 1973, the 246GTB engine was therefore replaced by the 308GTB unit, which was then considered a powerful new 2.9-liter V8 engine.

Within two years Ferrari were selling Bertone-bodied 2+2 seaters, and Pininfarina-styled two-seater coupés, the latter of which was known as the 308GTB. In 1977, the 308GTS Spider, which was the successor to the open-topped Dino 246GTS Spider, was also launched.

Both models could reach 158 mph (254 kph) and accelerate from 0 to 62 mph (0 to 100 kph) in 7.3 sec., although the Spider was slightly more expensive than the coupé. Their 179 ci (2,926 cc) engines developed 240 bhp at 6,600 rpm, thus earning them the official designation of Quattrovalvole, or QV.

The Ferrari Mondial QV also carried the same engine as the 308GTB, but reached a slightly less top speed of 149 mph (240 kph)

although had an acceleration of 0 to 60mph (0 to 97kph) in 6.4 sec.

When the model was first launched, 308GTB/GTS was considered a refinement of the previous 246GT/GTS type, with less pronounced wheel arch bulges, and more angular lines around the tail.

As with the earlier Dino, the GTS Spider was really no more than a GTB with a removable roof panel – especially since the screen and the rear quarter styling were retained for both cars. The GTS, like the GTB, had wind-up windows and even with the top removed, was a draught-free car at very high cruising speeds.

Below **The Ferrari 308GTB, styled by Pininfarina, was originally introduced in 1975 with a glass reinforced plastic body, which within a couple of years, reverted to a metal shell.**

Ferrari 328GTS

V8 | 270 bhp | 153 mph (246 kph) | 1980s

The 308-328 Ferrari series used the same basic chassis as the Dino of 1967 – 1973, but were fitted with a newly designed 90-degree V8 engine.

Ferrari had to fight hard to keep its cars abreast of American exhaust emission laws, one result being that the V8 engine was given fuel injection in 1980, and heads with four valves per cylinder in 1982. Following the success of these two changes, the engine was then enlarged to 195 ci (3,186 cc).

The result of all these changes was that 12 years after the original 308GT engine was announced, the engine had grown, and was 15 bhp and 15 lb ft of torque sharper than before.

Ducts into the engine bay, positioned ahead of the rear wheels, were fed by air along the flanks through sculptured channels in the door pressings, while a low and wide nose channeled air into the front-mounted radiator. At the rear of the car, the tail was sharply cut off, much after the style of the legendary Daytona. The sail panels linking roof to rear quarters were arguably more elegantly treated than those of the Jaguar XJ-S.

The 308/328 V8 engine was as powerful, and reliable as any other Ferrari unit. It was originally the first engine from Maranello to use cogged belt drive for its camshafts.

By the late 1980s, even though it was not available with four-wheel drive, anti-lock brakes, or other high-tech developments, the 328GTS was still seen as one of the world's most desirable open-top machines.

Right **Ferrari's 90-degree four-cam V8, launched in 1973, was the very first V8 to be sold by the Italian company. It originally had two valves per cylinder, but by the early 1980s it had been given four-valve heads and fuel injection.**

Below **The 308/328GTC models were extremely popular in the United States – especially in California – but were less common in Europe when they were first being produced.**

Ferrari Daytona 365GTS/4 Spider

V12 | 352 bhp | 174 mph (280kph) | 1960s – 70s

Ferrari built a series of stunning front-engined super cars in the 1960s, the most outstanding of which was the famous Daytona. Like most of Ferrari's two-seaters, this car was first conceived as a fixed-head coupé, but a Convertible was later developed.

Even by Ferrari's standards, the Daytona was a milestone; the ultimate expression of the big front-engined V12 bloodline, and the pinnacle of a

great Ferrari tradition. Yet it was also a docile and civilized Grand Touring car with a restrained elegance.

The designation 365GTS/4 represented the capacity of one of its dozen cylinders, the body type Gran Turismo Berlina, and the number of its cams – four.

This was the first ever Ferrari road car to have independent suspension at front and rear. Furthermore, it was distinguished by the use of a front-mounted V12 engine and a combined gearbox/final drive transaxle at the rear.

The car had a long and low nose, originally with headlamps hidden behind a transparent panel, a sharply raked windscreen, and a smoothly tapering roof over the two seats. One downfall was that luggage space was limited.

The Daytona not only looked magnificent, but had astonishing performance, road holding, and character. At the time, it was considered one of the fastest front-engined road cars ever built, which speaks volumes for its aerodynamic efficiency. A Daytona Spider with the soft-top down, quickly became idealized as the car for cruising along sun-drenched highways. It was, however, the last of its kind, for the front-engined Daytona was eventually displaced by the mid-engined Boxer, which was only ever sold as a closed car.

Left **For many this was the ultimate Ferrari convertible – for the 365GTS/4 was the last front-engined Ferrari V12 which also offered open-top motoring. The GTS was more rare than the fastback GTB/4, and in its day was considered much more valuable as well, largely due to its styling and performance.**

Ferrari Dino 246GTS

V12 | 352 bhp | 174 mph (280 kph) | 1970s

The first sports-racing Dino appeared in the 1960s. It had an enlarged iron-block V6 engine, and a slightly longer wheelbase.

Because of its mechanical layout—the engine was tucked in very close behind the seats—the 246GTS could not be a full convertible.

Instead, Ferrari, in conjunction with Pininfarina, copied Porsche's "Targa" idea, merely giving the car a removable roof panel, and the option of stretching out a soft top in order to keep out the rain.

The Dino was the first V6-engined Ferrari road car, the first to use a mid-engined layout, and was also the smallest-engined Ferrari for many years.

But most important of all was that it was not officially a Ferrari at all, but a Dino, for the cars were not sold with Ferrari badges.

Below **The Dino's engine had an impeccable pedigree designed by Jano for racing, productionized, and produced in numbers for Fiat, and used to power Ferrari and Fiat Dinos. The V6 engine had four overhead camshafts, and produced plenty of torque and power. From any angle, the Ferrari Dino was a gorgeous small two-seater; nobody could fault the Pininfarina lines. The Spider version kept almost all of the same lines as the coupé; the only "open-top" area was above the seats.**

Ferrari Mondial QV

V8 | 240 bhp | 149 mph (240 kph) | 1980s

The Mondial model came in two similar body styles, and carried the same engine as the 308 Spider. The car will accelerate from 0 to 60 mph (0 to 100 kph) in 6.4 sec, which was slightly quicker than the 308 Spider's acceleration. As with the 308 Spider, the open-topped Mondial Cabriolet was more expensive than the coupé. The cockpit combines leather-trimmed luxury with straightforward, strictly functional controls.

***Above* The Mondial Cabriolet was known for providing "wind-in-the-hair" motoring.**

Fiat 1500/1600 Cabriolet

4-cylinder | 58/72/80/90 bhp | 90 – 105 mph (145 – 169 kph) | 1960s

Although Fiat was Italy's most important car manufacturer, it did not seriously begin to develop sports cars until the 1950s. Its first attempt, the Trasformabile of the mid-1950s, was a fairly ungainly machine, and it was not until 1959, with the Pininfarina-styled Cabriolets, that Fiat found the correct balance in a sports car.

The new-generation Cabriolet was previewed in 1958, but did not actually go on sale until 1959. It was initially sold with a 1.2-liter pushrod Fiat engine, and a specially designed 1.5-liter twin-cam alternative, which was designed by the company OSCA.

The model then remained in production for seven years, during which time the pushrod engine was exchanged for a new 1.5-liter engine in 1963. The CISCA unit was slightly enlarged in the autumn of 1962.

The twin-cam cars gained front-wheel disk brakes from 1960, while the pushrod-engined cars gained front disks in 1963, at the same time as the twin-cam cars gained four-wheel disks. A new five-speed gearbox was fitted from 1965.

The Cabriolet had smooth lines with a characteristic front grille, uncluttered sides, neat tail-lamps on the rear corners, and wind-up windows in the doors.

Right **Pininfarina produced a simple, but exquisite, body style for the Fiat Cabriolets of the early 1960s. The same shape covered three distinctly different engines in seven years.**

Although the Fiat Cabriolets sold much faster than the Trasformabile had done, they were eventually overshadowed by the 124 Sport Spiders which took over from 1966. They were, nevertheless, the foundation upon which the 124 Spider's success was based.

Right **This Cabriolet interior, which dates from the early 1960s, was typical of Italian style of the period, with a neat instrument display, with much painted metal to back it all.**

Fiat Uno Turbo

4-cylinder | 105 bhp | 125 mph (201 kph) | 1980s

Fiat's unexpectedly rapid Uno Turbo model was launched in Rio during the weekend of the Brazilian Grand Prix in April 1985.

It had a turbocharged 1.3-liter engine with water injection and could accelerate from 0 to 62 mph (0 to 100 kph) in 8.3 sec. It could also claim an overall fuel consumption of some 32 mpg.

Uno Turbo belonged to the growing ranks of small, really fast cars. With the exception of this, really fast Italian cars were invariably exotic and very expensive. However the high cost of these cars ensured not only high performance, but more importantly, exclusivity.

The turbocharged engine gave the hatchback a top speed of 125 mph (201 kph). Despite the imposition of speed limits, Italy retains its reputation for making very fast cars.

Below **The boxy Fiat Uno Turbo may have looked like a fish out of water in the hallowed company of super cars, but the mixture of small hatchback and turbo power gave sports car drivers in the 80s a pleasant surprise.**

Fiat XI/9

4-cylinder | 75 bhp | 100 mph | 1970s

The two-seater Fiat X1/9 was revealed as a production model in November 1972. It had a distinctive wedge shape and a removable Targa top which could be stowed in the front luggage space. The fuel tank was located behind one seat and the spare wheel was behind the other.

The Fiat X1/9 used a mildly modified 128 engine, fitted more upright than in the sedans for better access. In single carburetor form, the 79 ci (1290 cc) overhead-cam 4-cylinder unit gave 75 bhp at 6000 rpm and 721b ft of torque at 3400 rpm – but this was reduced to a meager 66 bhp and 681b ft by American emission regulations.

Nevertheless, when it went on sale in the United States in 1974 it became a major success, undoubtedly due to its remarkable handling and road holding as well as its sheer style.

On 128-based MacPherson strut suspension, it gave almost neutral, fool-proof handling, instant steering response, and cornering power far beyond the reach of its mass-produced rivals, along with an acceleration of 0 to 60 mph (0 to 97 kph) in 13 sec.

Fiat was in no hurry to change this reputation – in 1976, it installed bigger bumpers for the United States, and in 1977 a right-hand-drive X1/9 was added.

In October 1978 the new Ritmo's 91 ci (1498 cc) engine provided the brilliant chassis with more power. The X1/9 1500 had 85 bhp and with improved torque and a 5-speed gearbox it also had a more flexible performance and up to 110 mph (177 kph).

During the 1980s, Fiat's quality control problems brought a fatal slump in sales for this car.

***Left* XI/9 styling is so neat that it is difficult to spot the mid-engine mounting. This is the 1500 5-speed model, with the bigger front and rear bumpers, and the raised engine lid.**

Ford (SA) XR8

V8 | 200 bhp | 140 mph+ (225 kph+) | 1980s

In July 1984, Ford of South Africa announced a variation on the XR4i theme, the XR8.

This was a limited edition of the Sierra model but with the Mustang V8 engine, rated at 200 bhp, plus a 5-speed close-ratio gearbox and disk brakes which were fitted all round.

The XR8 was intended as a homologation special and was a hybrid of the European Sierra body shell and American Ford V8 power.

The 5-liter engine ran the car up to more than 140 mph (225 kph) and from 0 to 62 mph (0 to 100 kph) in less than 8 sec. The car had the five-door body shell and was offered only in white.

When it was first being produced, Ford had intended to make only 250 cars, and was expecting to sell the full stock at a substantial price before the last one was finished being produced.

Right **The South African-built Ford XR8s was a hybrid of the European Sierra body shell and American Ford V8 power, and aimed unashamedly at putting Ford's South African cars in the forefront of production sedan racing.**

Ford Capri 2.8i

V6 | 160 bhp | 130 mph (209 kph) | 1960s – 80s

The Capri was originally introduced in 1968 and was immediately launched into a long and successful career. It was then a good-looking car and performed well for its time – especially in the old Mitre version and in the 2.8-liter model.

The ultimate Capri in Europe was the Injection Special – a stylish, hatchback coupé, that was well-made and finished. It delivered 160 bhp from its Bosch K-Jetronic injected engine, enough to hit 125 mph (201 kph) in fourth gear at 5,700 rpm. Acceleration from 0 to 60 mph (0 to 97 kph) was very brisk at 7.9 sec.

The Capri also offered Recaro seating, a leather-bound steering wheel, alloy road wheels, and a limited slip differential. Fuel economy, which is important in any car, worked out to an average of 25mpg, which was considered at the time to be good, especially in view of the car's performance. Riding on 205/60 VR 13 Goodyear NCT tires, the Capri was a good car to drive fast on dry roads. Its basic oversteer characteristics also allowed competent drivers to safely explore really fast cornering.

In wet weather, however, care had to be exercised, because as is the case with any front-engined, nose-heavy car, the tail can move out too far if the driver is careless. Braking was initially good, but from really high speeds stopping could be ragged.

Left **Ford were not well known for their production of impressive sports cars, but the 2.8 injection version of the enormously popular Capri coupé offered drivers exciting performance without breaking the bank.**

Ford GT40

V8 | 350 bhp | 165 mph (265 kph) | 1960s

In 1963, Ford, with a lot of help from the British racing car manufacturer Lola, began to develop the GT40, and in April 1964 the first car was shipped to the United States. It had the 256 ci (4,125 cc), all-alloy, pushrod V8 from the 1963 Indianapolis program, giving about 350 bhp and a theoretical maximum speed of close to 200 mph (322 kph).

By early 1965, the GT40 had grown into the Mark II, with 427 V8 power derived from the NASCAR Fords, detuned to only 427 bhp for endurance racing. The later Mark III switched the right-hand gear change

Above and Left **The GT40 was developed with one aim in mind—to win the Le Mans 24 Hour Race and beat Ferrari. And win they did, not just once, but four times in a row. Roadgoing GT40's are little different to the racing versions – with a 5 sec. 0 to 60 mph (0 to 96 kph) performance time.**

to a central lever and the excellent, fixed "hammock" seats and adjustable pedals for moveable seats.

The power and acceleration in the GT40 were truly awesome, with 0 to 60 mph (0 to 96 kph) in around 5 sec. and 100 mph (160 kph) in not much more than 10 sec. The car can achieve a top speed of 165 mph (265 kph). Even with the slightly softer road suspension and on wide wire wheels, the amount of grip and precision of handling were exceptional. The steering was considered quick, the gearbox slick perfection, and the only thing that used to betray the car was the massive effort needed to get the best from the big disk brakes.

The roadgoing GT40 was a very special legacy of "total performance" – known as a quick road car, and also the Ford that beat Ferrari.

Ford Merkur XR4TI

4-cylinder | 170 bhp | 125 mph (273 kph) | 1980s

The Merkur is a Sierra-based coupé, a three-door hatchback with the "biplane" rear spoiler.

Its engine is a 2.3-liter, 4-cylinder, single overhead cam unit with a Garrett TO3B turbocharger and gives 170 bhp at 5,200 rpm in manual gearbox form.

The car is front-mounted and drives the rear wheels via a 5-speed manual or Ford C3 automatic box. Suspension is fully independent and brakes are disks at the front and drums at the rear. Steering is by power-assisted rack-and-pinion. Alloy wheels carry 195/60HR 14 Pirelli P6 tires.

Below and Right **With the introduction of the "world car" concept, Europe and America have shared more ideas than ever before in order to foster international standards of car production. The Ford Merkur XR4TI was the American incarnation of the Sierra, with a turbo for good measure.**

Ford Mustang

6-cylinder | 277 bhp | 160 mph (257 kph) | 1970s

Lee Iacocca, also known as the Father of the Mustang, saw the need for a simple, relatively small, but fast and sporty car to sell to the growing number of young drivers.

By American standards, not only was the Mustang cheap and cheerful, but it was sold in an astonishing number of versions. The meekest of all had 6-cylinder engines, the most

Above **By the early 1970s the Mustang had become bigger, leaner, but not faster. It was, nevertheless, a very attractive soft-top.**

Left **The original Mustang was one of Ford's major triumphs, but the name was later used on some very ordinary cars. The latest Mustang GT goes some way to restoring some of the name's former reputation.**

extrovert had vast, rumbling, gas-guzzling V8s. Some had hardtop bodies, some were fastbacks, and a great number were convertibles. All of them were four-seaters.

Ford's in-house stylists had carefully contrived the rakish four-seater, on a 108 in. (274 cm) wheelbase, to look like a sporty two-seater. Each of these cars remained faithful to the long bonnet/short tail style of the original. The Mustang II which took over for 1974 was a much smaller, slower, and less exciting car.

Ford Sierra XR4i

V6 | 150 bhp | 130 mph (209 kph) | 1980s

The Sierra XR4i could perhaps best be thought of as a four-seater version of the Capri in coupé form. As with the Capri, a front-mounted pushrod V6 engine was used to drive the rear wheels via a 5-speed manual gearbox. The brakes are ventilated front disks and rear drums – also like the Capri.

The Sierra's suspension is, however, all-independent. Because of the car's more aerodynamic shape, it slipped through the air much better

than the older Capri, and the lower horsepower actually produced better acceleration, with 0 to 60mph (0 to 96 kph) in 8.4 sec.

Over long distances the XR4i's quieter mode was considerably less tiring than the Capri. Lacking a limited slip differential, however, and on narrower 195/60 VR 14 Goodyear NCT tires, the XR4i was not so pleasant to drive fast as the Capri. The Sierra's braking had a better feel than the Capri's stoppers, but on the downside the brakes tended to be noisy when used hard.

***Below* The Ford Sierra's styling was nothing if not controversial for its time; whether loved or hated, it had been affectionately dubbed "the jelly mold."**

Ford Thunderbird

V8 | 202 bhp | 115 mph (185 kph) | 1950s

Like the Corvette which preceded it, and the Mustang which was to follow in the 1960s, the Thunderbird was a stylish new sporting car which drew heavily on existing corporate parts already used in other models.

The Thunderbird, however, had a new 102 in. (259 cm) wheelbase chassis frame, allied to a smart new body style which, in the American fashion of the day, had barrel sides, a straight through wing line from headlight to tail-lamp cluster, and a fully wrapped-round windscreen with a pronounced dog-leg pillar.

Detachable spats covered the rear wheels, and there were styled air vents

in the tops of the front wings and on the bonnet panel. Unlike most European sports cars, the Thunderbird had a bench seat.

Although the running gear came from other Fords, with several different overhead valve V8 engine options, and with a choice of manual or automatic transmissions, the styling was at once different and refreshingly crisp. All these first-generation Thunderbirds were two-seater convertibles with capacious trunk accommodation, and there was a detachable hardtop available as an option. This top, which turned the car into a wind-and waterproof two-seater sedan, was plain-paneled at first, but from 1956 there was the option of circular portholes in the rear quarters.

The best of the Thunderbirds was undoubtedly the 1956 model.

Left **The last of the two-seater Thunderbirds was built in 1957, after which Ford produced a much larger, less sporty four-seater.**

Below **Latter-day enthusiasts can always identify a late-model two-seater Thunderbird by the "porthole" option in the detachable hardtop. This became available in 1956, but just to confuse those enthusiasts, some earlier cars were treated to the porthole top by their proud owners.**

Hispano Suiza Alfonso

4-cylinder | 65 bhp | 75 mph (120 kph) | 1910s

If the American Mercer Raceabout could stake a claim to be the first sports car in the world, the Hispano-Suiza Alfonso, introduced shortly after the Mercer in 1911, had an equally strong claim as the first production sports car in Europe.

It was designed by the great Swiss engineer Marc Birkigt, Hispano-Suiza's young technical director, and based on his successful voiturette racing design of 1909. A long-stroke 2.6-liter development of that car won the Coupé de l'Auto GP des Voiturettes in 1910. The first of several Hispanos to bear the Alfonso name emerged the following year, as a production version of the voiturette, with a side-valve four-cylinder engine, further enlarged to 3.6 liters.

It became available in a number of two-and four-seater versions, and they took their name from Hispano's enthusiastic royal patron, King Alfonso XIII of Spain.

By the heavyweight standards of the day, the Alfonso looked almost frail, with a high chassis, delicate looking wings, and narrow, wire-spoke wheels. The 3.6-liter engine was notable mainly for its very long stroke – more than twice the bore size and a feature originally prompted by racing rules. It was installed in unit with the gearbox, unlike most of its contemporaries which still tended to treat the latter as a separate component.

Even in 1911, there were faster cars than the Alfonso, but only either out-and-out racing cars or a very few large tourers. Where the Alfonso really differed was in its nimbleness and, like the Mercer Raceabout, in its style.

Above **Delicate-looking bodywork disguises a rugged chassis design. Its lightness gives the car good handling, and underneath lies a powerful racing-derived, 3.6-liter engine.**

Honda Acura NSX

V6 | 270 bhp | 168 mph (270 kph) | 1980s

In creating the first legitimate Japanese contender to the European super cars, Honda gave birth to the Acura NSX with all of the attributes of a Ferrari, and none of the vices.

To meet the weight stipulation, Honda used aluminum alloys just about everywhere on the NSX. On the suspension the NSX used a double wishbone design, with coil

springs and anti-sway bars. At the rear, an additional lateral link was used for better wheel control.

Brakes were vacuum-assisted disks all the way around, and used steel rather than aluminum calipers. An anti-lock system was used on the NSX, as well as an upgraded four-channel system that monitored each wheel independently using a 16-bit microprocessor.

The real key to the power Honda managed to unlock from its V6 is an extremely advanced camshaft/valve actuating system called VTEC – Variable Valve Timing and Lift Electronic Control System. The point of the VTEC system was to provide low-end torque and high-end power with none of the usual drawbacks.

A canopy effect was created for the NSX body, and the cockpit was almost a bubble that was positioned far forward in the body. To enhance that look even further, the canopy body panels were painted black to attract the eye to that part of the car.

In front of the cockpit was a short, wide, sloping nose that used pop-up headlights to increase the front rake. A small, split mouth was the main feature of the plastic front end cap, which boasted recessed parking lights and indicators, and a chin spoiler. Another refinement that Honda added to the NSX was the Traction control system.

Left **Acura's NSX was the first true super car to be built by a Japanese manufacturer. It combined civility with an aggressive design. The use of popup headlights allowed designers to keep the bonnet line low.**

Jaguar E-Type

6-cylinder and V12 | 265–272 bhp | 150 mph (241 kph) | 1960s–70s

Above **After ten years, Jaguar redesigned the E-Type, turning it into Series III with a brand new V12 engine. From this point, all types were built on the longer wheelbase underframe.**

Original E-Types had 3.8-liter 6-cylinder engines, but for 1965 the unit was enlarged to 4.2 liters. Then, in 1971, the E-Type became the first Jaguar to use the brand new 5.3-liter V12 engine.

Structurally, and in its styling, the E-Type was a further evolution of the D-Type/XKSS layout, with a monocoque center section, a multi-tubular front end, and the same sleek and flowing lines which were never successfully copied by any other car maker. The first E-Types had headlamps tucked away behind transparent covers, as well as high and rounded tails and rather minimal two-seater accommodation.

The demand for soft-top roadsters and fixed-head types was almost equally matched, the fixed-head model being arguably better styled. The Roadster, on the other hand, still had wind-up windows in its doors and a full-size curved screen, plus a folding soft top which clung to its job even at very high speeds.

Below **In the 1960s, the Jaguar E-Type became one of the most easily recognized cars in the world – all had curvaceous rears, with the exhaust system taking a prominent part in the style.**

The earliest cars were quite cramped; they tended to overheat; the faired-in headlamps were hopeless; the all-disk brakes were marginal for such performance; and the early gearboxes were dreadfully slow and heavy. Jaguar overcame most problems very quickly. They cured many faults simply by going into proper tooling. The car had great handling and brakes, as well as an excellent engine.

When it was eventually dropped in 1975, the world of motoring grieved. And even for Jaguar, the original E-Type was a very hard act to follow.

Jaguar SS100

6-cylinder | 125 bhp | 101 mph (162 kph) | 1930s

Jaguar's celebrated history began in the British seaside resort of Blackpool with the Swallow Sidecar Co., which built motorcycle sidecars in the early 1920s.

The company's first real sports car appeared in March 1935: a rakish two-seater with a 2.7-liter, 86 bhp, 6-cylinder Standard side-valve engine in a shortened SSI under-slung chassis. It was dubbed the SS90, alluding, slightly optimistically, to its top speed.

Only 23 SS90s were built up to September 1935 – when SS announced two new cars with overhead-valve engines – the SS Jaguar sedan and the SS Jaguar 100. The SS100, with its sweeping wings, wire wheels, and huge, mesh-covered

headlamps, looked very much like the two-seater SS90 but was much better developed. It used a purpose-built chassis rather than a cut-down sedan type, with an alloy body over an ash frame.

The overhead valve conversion and twin carbs put the quoted power output up to 100 bhp and the "100" again represented the claimed top speed. It stayed in production until 1939 and a total of 198 were eventually built, but toward the end of 1937 the car was finally able to live up to its name when a 3.5-liter version was introduced alongside the 2.7-liter.

The bigger engine, with slightly higher compression, offered great flexibility and the sparkling acceleration took it to 60 mph (96 kph) in about 11.5 sec. Its road holding and handling, with very quick steering, were impeccable – even if the ride was a bit hard – and its brakes were superb. Just 116 3.5-liter SS100's were built before they too ceased production, in 1939.

Left **Performance and style at a budget price? The classic SS100 Jaguar sports car is a far cry from the company's original products – motorcycle sidecars. Faultless handling, superb brakes, and a willing, flexible engine made the SS100 a real driver's car during the 1930's.**

Jaguar XJ-S Cabriolet

6-cylinder, V12 | 295 bhp | 141 and 153 mph (227 and 246 kph) | 1980s

For the first eight years of its life, the XJ-S was only available as a close-coupled four-seater with fixed-head coupé styling and the powerful but gas-guzzling V12 engine.

It was not until the autumn of 1983 that Jaguar produced a cabriolet version of the XJ-S, and at the same time a brand new, twin-cam, 6-cylinder engine. The XJ-6 unit was also put on sale. This was mainly intended for the XJ40 sedan car, which appeared in 1986, and was improved in detail during its trial run in the XJ-S model.

Because the XJ-S was based on XJ-6 sedan-car engineering, it was a solidly proportioned car built around pressed-steel panels, with impressive ride and handling.

The V12 engine, although magnificent and powerful, was also

Right **Below the waistline, and ahead of the screen, the open-top SC looks the same as the XJ-S Coupé. Only the roof and the controversial 'sail' panels have been discarded.**

very large, and even in "HE" form it was quite thirsty. But it was without a doubt, one of the most silent big engines in the world during its time.

The cabriolet of 1983 used the same basic body lines as the coupé, except that the entire top and rear of the body were reshaped. The sail panels disappeared completely and the accommodation was cut down to two seats, yet the car retained fixed rails around the door glasses, as well as a strengthening brace across the car between the pillars, above and behind the passengers' heads.

A Targa panel in the roof itself was, of course, removable. The soft-top itself was well-padded and tucked down into the area previously reserved for the rear seats of the coupé. Luggage accommodation was as generous as that in the coupé model.

Left **Because of the rather bulky fold-back soft-top fitted to the XJ-SC, there was no space for rear seats. The XJ-S was a generous 2+2 model, whereas the XJ-SC was a two-seater.**

Jaguar XJ-S HE

V12 | 299 bhp | 155 mph (249 kph) | 1980s

Production of the XJ-S actually ceased for a while, but on arrival of the HE engine sales gradually increased. If it is a two seater or 2+2 Jaguar that is required then the XJ-S HE model is an ultimate classic. It was no sports car, but it served as an admirable executive's express. Equipped with a fuel-injected V12

Right **Jaguar purists may never have been quite sure about the XJ-S's controversial shape but this car, with its superb V12 engine, was quite simply one of the best cars of its times.**

engine, that even Mercedes-Benz claimed was the best mass-produced high-performance engine of its time, the big coupé had no real rivals in its homemarket during the 1980s.

The 5.3-liter overhead-cam engine uses Bosch digital electronic fuel injection and an electronic ignition system.

Fuel economy works out at 22.5 mpg at 75 mph (120 kph), which is quite remarkable for this large, very fast car which will storm from 0 to 60 mph (0 to 96 kph) in 7.5 sec.

Beautifully made and finished, this car was hugely popular in the 1980s and Jaguar found it hard to meet customer demands.

Jaguar XJ-6

6-cylinder | 205 bhp
130 mph (209 kph) | 1980s

The XJ-6 4.2-liter is powered by an in-line twin-cam engine, possibly the final flowering of the same engine that powered the XK-120 over 30 years ago.

It uses Bosch L-Jetronic fuel injection to produce 205 bhp at 5,000 rpm, enough to accelerate the car from a standstill to 60 mph (96 kph) in 9.8 sec.

As with all the XJ models, that is not the whole story. On a value-for-money basis, nothing touches the Jaguar. Smooth and quiet with a silky quality that nothing else can match, the car from Brown's Lane, Coventry put every other high-quality fast car into the shade.

Until the arrival of John Egan, as managing director, Jaguar suffered from a build-quality problem, but under his direction this changed dramatically, and developed waiting lists. Buyers were snapping up all models of the car worldwide.

The Jaguar revival was particularly apparent in both North America and Germany.

Right **Unmistakably Jaguar: the simple lines of the XJ-6 are instantly recognizable.**

MICHELIN

Jaguar XKSS

6-cylinder | 250 bhp | 144 mph (232 kph) | 1950s only

Jaguar, having launched the famous twin-cam XK engine in 1948, soon got involved in sports car racing, first with modified XK 120s, and from 1951 with the specially developed XK 120C. Then, in 1954, Jaguar produced the racing D-Type, initially as a works racing car, and from 1955 as a limited production machine for worldwide sale.

But more D-Types were built than could be sold, so at the end of 1956 it was decided that the unsold stock of D-Types would be slightly reworked and sold as "road-going" cars called XKSS.

For the XKSS conversion the D-Type was given a full-width curved windscreen, the headrest and the metal spine between the body seat cut-outs were removed, a second door was added to the left side, and fixed side screens were added to those very steeply contoured but tiny doors.

Smart bumpers were added to the front and rear, a luggage rack was fixed to the top of the tail and a fold-down soft-top was also added. Because the entire front end was hinged at the nose, there was excellent access to the engine bay and front suspension for maintenance work.

The XKSS, like the D-Type, looked beautiful because of its elegant and well-curved styling. Most of the running gear was adapted from that used in Jaguar road cars, which explains why so many replicas of this car have been constructed.

Right "The office" of the XKSS was slightly more civilized than that of the D-Type, but still had restricted leg room, and virtually nowhere to stow anything. But with all that performance in reserve no-one seemed to care.

Below When Jaguar converted D-Types into XKSS road cars, it discarded the hump behind the driver's head, added a full-width screen, and a fold-down soft-top.

Koenig GmbH

Depends on car | Custom | Custom | N/A

At the top of Koenig's range is their reworking of the big coupé, the 500SEC. During the early 1990s, this model was popular with businessmen and film stars, and the first Koenig SEC completed was for Sylvester Stallone.

Sly's SEC was equipped with a supercharger, boosting the 5.6-liter V8 engine. Bulbous wheel arches over massive tires and a deep front air-dam balanced the spoiler on the trunk, and the interior was completely retrimmed in leather.

Even the SEC conversion, however, was surpassed by another Koenig offering. They started with an already outrageously fast Ferrari Testarossa and turned it into a turbocharged rocketship. In its "basic" version, the Koenig's twin turbochargers increased the output to 750 bhp.

For the truly courageous, the tuning company created yet another option: they fit even bigger turbos to the model. Around the large control knob the figures backed up Koenig's claim that this was one of the most powerful cars on the market.

Astonishingly, there was a faster car in Koenig's range. For this adventurous tuning and conversion company also had its own version of Porsche's 962 sports-racer – the C62. At a phenomenal top speed of 235 mph (378 kph), the C62 relied on ground effects to keep it in contact with the tarmac, and its flat-six, turbocharged engine hurtled the car past 60 mph (96 kph) in 3.5 sec.

Right **BMW's luxurious 850i coupé, widened and dropped almost to ground level by Koenig. It was impractical outside the racetrack, but some clients were willing to suffer for style.**

Below **Mercedes designed their SL sports car to be handsome and Koenig made sure you couldn't miss it. With its low suspension, wide wheels, extended arches, and a hefty tail spoiler, it's unmistakable.**

Lamborghini Countach

V12 | 445 bhp | 190 mph (306 kph) | 1970s

According to legend, local workers gave the name for this Ferrucio Lamborghini super car. They took one look at the sleek, futuristic, swing-up door device and exclaimed, simply: "Countach!" – the nearest translation is something to the effect of "Wow!"

Although it uses essentially the same V12 engine as the Miura, the Countach had a more conventional arrangement with its 5-speed gearbox on the forward end of the block, bringing the gearbox between the occupants and the gearlever to the driver's hand with no additional linkage. Less conventional was the drive which is fed back by a shaft running through the sump, to a final drive at the back of the engine.

The Countach did not go into proper production as the LP400 until 1974, by which time it had undergone several changes from the prototype. The engine had returned from 5-liters to the Miura's 4-liters, and the planned monocoque chassis gave way to a multitubular space frame, with welded-on alloy panels. It retained the dramatic door opening arrangement by which the doors swung up and forward in an arc around their top leading corner, counterbalanced by hydraulic struts.

The fully adjustable suspension had coil springs and wishbones at the front, and coil springs, wishbones, top and trailing links at the back. The brakes too, were virtually to racing

Right **Little more than a thinly disguised racing car, the Lamborghini Countach offers stunning performance. The sleek alloy bodyshell hides a tubular spaceframe chassis. Rear visibility is negligible through the tiny rear screen.**

specifications, with large ventilated disks and four-pot calipers.

Changes began in 1978 with the much better shod Countach 400S, which also had considerably wider wheel arches, plus the option of a large rear aerofoil – giving the car a more aggressive look. The next change came in 1982 with the Countach LP500S, which had a 4.8-liter version of the V12, with a claimed 400 bhp.

As a performance car, the Countach wasn't very comfortable. It was also noisy and uneconomical when it came to petrol consumption.

Lamborghini Diablo

V12 | 495 bhp | 202 mph (325 kph) | 1970s

Constant development has seen the capacity of the Diablo rise to 5.7 liters, with four valves per cylinder, fuel injection, and an impressive 495 bhp. Power like this catapults the Diablo from 0 to 60 mph (0 to 96 kph) in 4.2 sec. and later to 100 mph (160 kph) in 4.3 sec.

The Diablo engine drives forward to a clutch and 5-speed gearbox by the driver's elbow, from where a transmission shaft runs back through the block under the crankshaft to a differential just behind the engine.

Apart from spreading the weight distribution evenly and bringing more weight to the front of the car, it also means that the gear-lever engages the shift mechanism directly without the need of shift rods in the box. This reduces weight and produces a lighter, swifter change. It also has another big

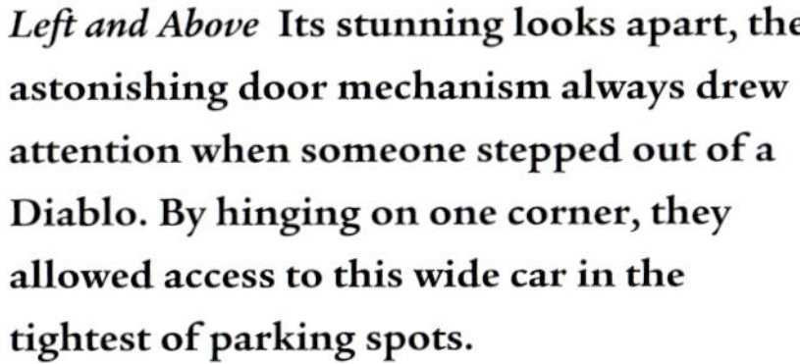

Left and Above **Its stunning looks apart, the astonishing door mechanism always drew attention when someone stepped out of a Diablo. By hinging on one corner, they allowed access to this wide car in the tightest of parking spots.**

Above Right **With the gearbox between the seats, the stubby gear lever has a crisp positive feel. Lie back on the soft leather upholstery and the controls are ideally placed; high-speed travel demands a relaxed driving position.**

advantage – four-wheel drive.

Under its coat of deep gloss paint, the Diablo combines steel, aluminum, and Kevlar composite panels over its steel structure. Suspension under the Diablo consists of cast alloy double wishbones at each corner, with coil springs and firm dampers.

The specially developed Pirelli P-Zero rubber have three separate tread zones: on the inside, small blocks cope with wet conditions; a central stripe of slick rubber offers maximum traction and braking, while the fine tread on the outside shoulder gives immense cornering power on dry tarmac.

The other winning feature of the Diablo is the outrageous guillotine door, which pivots vertically from its front edge.

Lamborghini Miura

V12 | 350 bhp | 170 mph (274 kph) | 1960s–early 70s

In the 1960s, when Italian industrialist Ferruccio Lamborghini branched out into the hazardous business of building sports cars, he did so with the intention of beating Enzo Ferrari at his own game. And with the Miura he beat his rival into the mid-engined super-car league by eight full years.

Before the Lamborghini Miura, only racing cars were mid-engined. It had a unique transverse V12 engine layout, and its stunning Bertone-designed body boasted a simple beauty that has rarely been rivaled.

It started as the P400, with 350 bhp, and grew with appropriate chassis improvements through the

Below **Under the tilting tail, the Miura's unique transverse V12 engine layout is obvious. Suspension loads feed into the monocoque, leaving the "clam-shell" body sections unstressed.**

Left **With Italian engineering and styling, the superb lines of the Bertone-styled Miura embody wind-cheating aerodynamics that help it to a top speed of 170 mph (272 kph).**

370 bhp P400S of 1969 to the ultimate Miura, the 1971 P400SV, with subtle but unmistakable aerodynamic tweaks based on a stillborn racing design, the Jota.

With wishbone and coil spring suspension all round, it combined a supple ride with marvelous road holding – although it was undeniably nervous around its eventual limits, especially in the early versions, and it did have a tendency to lift its nose at very high speeds. It was also short on luggage space and the interior luxury was offset by a dozen big air intakes wolfing greedily away only inches behind the occupants' heads.

The Miura was a landmark; an engineering *tour de force* which rewarded a skilled driver with staggering performance and lesser mortals with a special symbol of style.

Lancia Aurelia Spider

V6 | 118 bhp | 108 mph (174 kph) | 1950s

The running gear and the styling of the Aurelia Spider were notable. The sedan naturally used a steel-paneled unit construction shell, but it was still possible for its underframe to be used as the basis of shorter-wheelbase sporty models. The body shape was sexily curvaceous and attractive, and used a wrapped-round windscreen.

The running gear featured a powerful V6 engine, along with a rear-mounted transmission, and the long-established Lancia sliding pillar front suspension with De Dion rear suspension.

Above **The Aurelia Spider's original style was by Pininfarina, who included several Detroit touches to the shape, including the wrapped-round windscreen. Later types were fitted with a more conventional screen shape, allied to wind-up door glasses. Lancia's shield badge fits proudly to the famous grille, the result being an aggressive but pleasing sports car body style.**

Lancia Stratos

V6 | 190 bhp | 142 mph (235 kph) | 1970s

Having been taken over by Fiat, Team Director Cesare Florio was able to select the powerful 2.4-liter V6 engine from the Ferrari Dino, also part of the Fiat Empire. He chose a mid-engined layout for traction, and installed it in a steel monocoque built by Bertone.

The fiberglass body was the most startling rally car ever unveiled, and enthusiasts cheered as it snarled its way to World Championships in 1974, 1975, and 1976.

Below **The road-going version of the mid-engined Lancia Stratos dominated the world rally scene for years and it looks like a one-off show special, but the amazing Stratos brought three world rally championship to Lancia. Nimble, light, and strong, its Ferrari engine makes it a thrill to drive.**

Lancia Thema

4-cylinder | 165 bhp | 135 mph (217 kph) | 1980s

After years of hovering on the edge of extinction, the name Lancia was again linked to a car with real performance pretensions – the turbocharged Thema.

Clean, modern, and almost anonymous, would be a good way to describe the Thema's styling but, as always with Lancia, it is what is under the bodywork that is important.

The engine of the fastest Thema is a 2-liter four-cylinder, turbocharged, fuel-injected unit delivering 165 bhp through a 5-speed gearbox. This engine has two counter-rotating balance shafts to smooth out the four-cylinder roughness.

The fuel injection is by the Bosch LE-2 Jetronic system, aided by Magnetti Marelli Microplex microcomputer ignition control, which features an "overboost" arrangement allowing the driver to select, for very short periods, a delay in the turbo waste gate opening which has the effect of increasing torque from a normal 188 lb ft to 210 lb ft.

Right **Lancia has rarely resorted to styling excesses to sell its car, preferring instead to rely on fine engineering – a policy which in the past has caused the company more than its share of financial crisis. The turbocharged Thema, styled under the direction of Giugiaro and the fastest Lancia for many years, is no exception.**

The ZF power-assisted rack-and-pinion steering has variable rate characteristics, lowering the ratio at low speeds, to help in parking maneuvers, and quickening up at higher speeds for more sensitivity. Self-leveling rear suspension and ABS braking are also offered on the turbocharged Thema.

Lancia claim a 0 to 60mph (0 to 96 kph) acceleration time of 7.1 sec. At speed the old Fiat 2-liter engine note is absent in the turbo car, but there is some wind noise.

Lexus SC400

6-cylinder | 277 bhp | 160 mph (257 kph) | 1990s

The car tops 150 mph (240 kph), generates nearly eight-tenths of a G in lateral adhesion, hits 60 mph (96 kph) in 6.5 sec. and does it all without rattling the driver.

At the root of the SC400's performance charm is the 342 ci (3,969 cc) V8 that is lifted mostly intact from the LS400 sedan. It is an all-aluminum creation that has four valves per cylinder, actuated by two overhead camshafts.

Fuel is delivered through a multiport electronic injection system, which is monitored by an anti-knock sensor that keeps the engine from developing a knock under acceleration.

The steering wheel, which contains an airbag supplementary restraint system, both tilts and telescopes electrically at the touch of a button on the left side of the steering column.

Seats are motorized for infinite adjustment, and electrically heated seat bottoms are optional. An additional feature is a standard seat memory system that allows two people to set their favorite seat/steering wheel position.

Above **Viewed from the top, the curvaceous nature of the Lexus coupé shows through clearly. There is not a single straight line in the SC400's design**

Left **The Lexus SC400 coupé is one of the most stunning designs of the 1990s**

Apart from the traction-control system, there is a rear spoiler, a power tilt-and-slide moon roof, and a remote-control Nachamichi compact disk sound system. Also available is a remote controlled security system that makes an SC400 very difficult to steal.

Lotus 7

4-cylinder | 135 bhp | 112 mph (217 kph) | 1950s

There was certainly no compromise about the Lotus 7. It was an out-and-out sports car. It has two seats and virtually no luggage space, rudimentary weather protection in the guise of an awkward-to-erect hood and side screens, and very few other creature comforts.

However, it does have a race-bred tubular chassis and suspension which contribute to road holding and easy handling.

A long-time Lotus dealer, Caterham Cars, in Surrey, England, acquired production rights from Lotus in May 1973 and has since been producing the car, with constant improvements.

With any of the more potent engine options it will nudge six seconds for the dash to 60 mph (96 kph) – even though its top speed is limited by the aerodynamics. Over the years, engine options grew from small pushrod units, through

Formula Ford-type units, the famous Lotus Twincam in various stages of tune, to that engine's modern replacement, the big-valve, twin-cam built for Caterham by Lotus preparation specialists Vegantune.

In 1982 Caterham added a long-cockpit option to give bigger drivers a touch more leg-room but neither that nor the myriad engine and chassis variations have changed the character of the Lotus 7 in the slightest.

Below and Left **Four-wheeled motorbike? The thoroughly individual Lotus 7 pays no heed to modern comforts. The basic design of the car hasn't changed since 1957 – it's fast, it's fun, and it's very impractical.**

Lotus Elan

4-cylinder | 105 bhp | 110 mph (177 kph) | 1960s

The Elan introduced the classic Lotus backbone chassis, forked at the front to carry the legendary four-cylinder twin-cam engine and coil spring and wishbone suspension, and less-markedly at the back to take a Chapman strut with wide-based lower wishbones and rubber-jointed driveshafts.

The Elan for the 1960s was a compact, fiberglass-bodied, drop-head, two-seater, powered by various versions of the Ford-based Lotus twin-cam engine. It was shown first at the London Motor Show late in 1962, but didn't reach production until 1963. Production then didn't necessarily mean complete assembly; a British tax concession of the time (or loophole) said that the cars bought as "kits" escaped purchase tax. Lotus was one of many specialist car makers to take advantage of leaving the wheels off and engine out, but was one of the few to survive when the concession disappeared.

The original 1.5-liter motor gave 100 bhp but only a few were made before the classic 105 bhp 1.6-liter unit was introduced, the original engines being recalled and replaced, free of charge. As ever, the bottom-line power grew over the years, with up to 126 bhp on offer from 1971 in Sprint versions. There were other changes, like a fixed-head coupé from 1965 and even a slightly bigger +2 sister model from 1967, but they were all Elans.

The fiberglass reinforced plastic body, with its neat pop-up headlamps, never changed much and the biggest first-glance differences between early and late Elans are gradually increasing tire widths and a hood bulge on the third series.

It is easy to feel at home in the Elan; functional design was Chapman's immutable creed. The car is just over 12 ft (3 m) long and knee-high, but it is amazingly comfortable even for the biggest driver. The stubby gearlever flicks from ratio to ratio just like a switch, the handling was as sharp as a kart on rails, and the all-round disk brakes simply stopped the car on demand. The balance between soft springs and firm damping gave a sweet ride in spite of the low weight and the actual grip was considered astounding, even around the tightest corners. It wouldn't do more than 110 mph (176 kph), but on an average road it ran rings round cars 50mph (80 kph) faster.

Above **Functional in design and layout, the Lotus Elan combines simplicity with leech-like roadholding and a willing (if occasionally unreliable) twin-cam engine.**

Lotus Esprit Turbo

4-cylinder | 210 bhp | 152 mph (245 kph) | 1980s

The 1985 Esprit Turbo had a number of all-new features, in particular a new chassis and revised front suspension, as well as new brakes, electrics, tires, and body and trim details. The chassis even came with an eight-year anti-corrosion warranty.

The heart of the Turbo was the mid-mounted 2.2-liter, 4-cylinder engine with four valves per cylinder. The turbocharger blew through two Dellorto carburetors which are mounted as close as possible to the intake manifold so that turbo-lag was virtually eliminated.

Acceleration from 0 to 100 mph (0 to 160 kph) took an astonishing 14.6 sec. Remarkable fuel economy bore out the efficiency of its engine.

The car was given a dramatic shape by the Italian stylist Giugiaro and like so many of his best designs it gave the impression of a fast car even when standing still.

Above and Right **The mid-engined Lotus Esprit Turbo amply demonstrated Lotus's late founder Colin Chapman's philosophy of "no bloody compromise." If you wanted this kind of performance you had to travel light and if you didn't want to hear the engine noise, buy a bike.**

Lotus Etna

V8 | 340 bhp | 182 mph (293 kph)
1980s

In October 1984 Lotus announced their new "Concept" car, called the Etna. Styled by Giugiaro at Ital Design, it had one of the cleanest, most dashing body shapes ever seen on a British car.

Slightly larger than the Esprit model, it had the exciting new Lotus V8 engine installed.

The engine was the jewel in the crown of the remarkable new Lotus, and the first production cars came off the Hethel production lines.

The whole car promised to be state-of-the-art in super-fast car construction and performance. It was by far the quickest and most sophisticated Lotus during the 1980's.

Lotus Excel

4-cylinder | 160 bhp | 134 mph (216 kph) | 1980s

The Excel had the performance to go with its good looks, with a top speed of 134 mph (216 kph) and 0 to 100 mph (0 to 160 kph) in 20 sec.

Lotus altered the model quite significantly, with new body styling, wheels and tires, electrics, instruments, air-conditioning, hard trim and soft trim. However, the

chassis was still the familiar steel backbone with a five-year anti-corrosion warranty.

Thanks to the restyled body, the rear window was 25% bigger and access to the trunk was much improved. New VDO instruments called Night Design were relocated in the dash panel to better effect.

The engine was the 912-type, 2.2-liter, 16-valve, double overhead cam 4-cylinder unit, which produced 160 bhp at 6,500 rpm. It had a 5-speed gearbox and rear-wheel drive.

All-independent suspension displayed all the features that Lotus was famous for, the very highest levels of road holding with absolutely no sacrifice to ride or comfort. Power-assistance to the rack-and-pinion steering was optional. Ventilated disk brakes were used all round and the system was servo-assisted. Alloy road wheels of 14 x 7 in. (35½ x 18 cm) carried 205/60 VR 14 Goodyear NCT tires. At the time, this car was a good all-rounder.

Above **The four-seater Lotus was big but not heavy, quick but easily handled, and exclusive but not unattainable.**

Left **A subtle rounding out of the Lotus Excel's extremities gave it looks to match the car's performance from the neat 16-valve 4-cylinder engine.**

Maserati Khamsin

V8 | 280 bhp | 160 mph (257 kph) | 1970s

The Khamsin model is the final embodiment of the great front-engined, high-speed Maseratis, like the Mexico and the Mistral.

The Khamsin's handsome and stylish body houses a front-mounted 287 ci (4,700 cc) V8 engine, with double overhead cams, driven by a duplex chain. This power unit breathes through four Weber 42DCNF41 carburetors.

The Khamsin body, although officially described as a 2+2, is really only a two-seater, as rear leg and head room are minimal at best.

The steering and seats are adjusted hydraulically, and the steering column can be adjusted for both length and rake by this system. The body styling is tasteful and the car is very well finished. It accelerates from 0 to 60 mph (0 to 96 kph) in 6.5 sec.

Right **Although the Berton-styled Maserati Khamsin, introduced in 1974, is now growing somewhat long in the tooth, it must still be regarded as a classic example of the traditional, big, front-engined supercar.**

Right **Series 1 Khamsins have one front grille under the front bumper, whereas Series 2 cars, as shown here, have a second front grille over the front bumper between the headlamp-covers.**

Maserati Merak SS

V6 | 208 bhp | 143 mph | 1980s

The Merak SS's engine was a 181 ci (2,965 cc) V6 with double overhead camshafts driven by a duplex chain. This engine produced 208 bhp at 5,800 rpm on three Weber twin-choke carburetors – two were 42DCNF31s and the third was a single 42DCNF32.

However, there was one drawback to this engine – on driving the Merak at maximum revs for any length of time, an alarming drop in registered oil pressure was noticed. This may have been due to the all-alloy engine flexing, allowing the main bearings to release pressure, with dire effects on engine life. The lack of a modem fuel-injection system showed up in temperamental cold starts and fussy driving habits until the engine was fully warmed up.

On the road, the Merak was stable, with a reasonable amount of understeer at all times, its ride was choppy at low speeds, improving with an increase in speed to become acceptable. Braking was by the Citroën high-pressure system and really wasn't well suited to such a high-performance car, as its oversensitivity hardly made for smooth progress.

The driving compartment was comfortable, with enough room for two average-sized adults, but the heating and ventilation systems were poor, taking a very long time to warm up, and then control of the heat was difficult and ventilation almost non-existent!

Right and Below **Somehow, the Maserati name has never had quite the same *cachet* as Ferrari, but the Neptune's trident badge of the city of Bologna has nonetheless graced some superb cars. The Maserati Merak SS was introduced in 1975, the year in which Citroën, who had rescued the Bolognese company in 1968, relinquished control to Allesandro de Tomaso. Under de Tomaso's management, Maserati made a remarkable recovery and for once in its troubled history could face the future with real confidence.**

Maserati Bora

V8 | 310 bhp | 157 mph (253 kph) | 1970s

In many people's eyes, the best Maserati of all came at a time when the company was going through an even deeper crisis than usual. This innovative new car came in the form of the V8-engined Bora, widely acclaimed but often overlooked in the super car stakes.

Like other new generation super cars, it was mid-engined, in this case with a truly excellent V8 of Maserati's own making and developed from their successful 4.5-liter sports racing cars of the mid-to-late 1960s.

It was a light, compact, all-alloy unit with twin overhead cams and fed by

four twin-choke downdraught Weber carbs. The Bora started life at 4.7-liters, which gave an impressive 340 lb ft of torque. It grew later to just over 4.9-liters, adding another 20 bhp and becoming even more willing at low revs.

On coil spring and wishbone suspension, the car had near-perfect handling balance and exceptional grip. And if it had a fault it was only in the braking system.

All this sat in a tubular subframe at the back of a steel unit-construction two-door coupé shell, by stylist Giugiaro. It was properly equipped, and even had automatically adjustable seats and pedals.

Below **The Maserati Bora, like many super cars of its day, is mid-engined and developed from Maserati's racing expertise. Its main drawback is the Citroën derived braking system which has little "feel."**

Mazda MX-5 Miata

4-cylinder | 116 bhp | 117 mph (188 kph) | 1990s

Mazda studied the characteristics of cars such as the MGB, the Triumph TR4, and the Lotus Elan, kept the good parts, threw out the bad, and created the MX-5 Miata – or just Miata for short.

The Miata engine fit snugly in a more conventional front-to-back position. It was a 1.6-liter in-line 4-cylinder that used twin overhead camshafts and four valves per cylinder. The engine was linked to the rear wheels by a 5-speed manual gearbox.

The Miata rode on a wishbone-design suspension that used low-mounted upper and lower A-arms at the front, along with moderately soft coil springs, gas shock absorbers, and an anti-roll bar. At the rear was a nearly identical set-up that used unequal bushing elasticity to generate a certain amount of toe-in under hard cornering.

The Miata was a convertible with just two seats and an overall jaunty air to it. With the help of stylists at Mazda's Southern California design studio, the Miata's shape was hammered out.

Below **Popup headlights gave the Miata a look that was similar to the Frog-Eye Sprite from the late 1950s.**

Above **A Miata pictured against the imposing shape of Hawaii's Diamondhead.**

Its engine was a fine, tightly built unit that produced enough horsepower to move the Miata along briskly, but its top speed was nothing to brag about. Its suspension was well engineered, but it used no computers or space-age alloys. Its cockpit was well appointed, yet it was small and there was not a computer display to be seen anywhere. Its styling was clean and handsome, yet also mild-mannered.

Mazda RX-7

16-valve | 116 bhp | 117 mph (188 kph) | 1990s

Where every other sports car in the world uses the traditional reciprocating piston engine, the Mazda RX-7 used the brilliantly simple Wankel rotary engine, an engine that offers wide-ranging benefits.

Dr Felix Wankel's engine, which basically comprised a three-sided rotor following a four-stroke cycle within an outer casing, was developed during the 1950s by Wankel and the German car and motorcycle maker NSU. The engine solved the problem of converting the energy of burning fuel directly into rotary motion.

The second generation RX-7 had new body styling, a clever new suspension system, and a 2.6-liter version of the trusty twin-rotor engine – giving 135 lb ft of torque at only 3,000 rpm for superb flexibility and smoothness.

The car was very well equipped for a sports car, with such standard fittings as an electric sunroof, electric

windows, an up-market stereo system, air-conditioning for the American market, and much more.

The new suspension was excellent. The front used MacPherson struts with lower A-arms, coil springs, and an anti-roll bar, but the real novelty was at the back, which had a multi-link system with floating hubs, coil springs, and an anti-roll bar – with telescopic dampers all-round.

The main feature of the "Dynamic Tracking" rear suspension was variable geometry, which allowed a small degree of initial toe-out during cornering changing to a degree of toe-in, which helped neutralize oversteer. Coupled with adjustable damping rates and superb variable rate power steering, it gave the RX-7 very high levels of grip, very precise responses, and even a reasonably good ride.

Below **Mazda's RX7 coupé was powered by the turbine-smooth, Wankel rotary engine. Only Mazda have persevered with the rotary, and it had finally paid off with a powerful smooth-running, twin-rotor unit that endowed it with performance akin to that of a Porsche 924S. Even more remarkable was that the rotary engine did this with an engine of just over 1.3 liters, yet delivered performance similar to a 2.4-liter engine.**

Mazda/Elford 929C Turbo

Rotary-turbo | 135 bhp | 127 mph (204 kph) | 1980s

The Elford Engineering Company in Great Britain turned their hand to the rather gutless Mazda 929 coupé, which shared the Wankel engine, giving it turbo-power and turning it into one of the best value-for-money four-seater cars in Europe.

Without the turbo, the car had a real struggle to get to 98 mph (158 kph), with the meager 90 bhp that is available from the four-cylinder single overhead cam engine. The Elford-Garrett T3 turbo changed all that.

In making the car faster, the extra power changed the function of the gearbox from a 4-speed plus overdrive fifth, to something that felt much more like a close-ratio sports box.

In top gear the car benefited from the added engine power (instead of struggling to maintain a fast cruising speed) and accelerated rapidly up to its 127 mph (204 kph) top speed with ease, and was able to maintain the speed in face of most gradients.

The turbo installation was neat, and looked like a factory fitment. The interior was equally good, with all the controls and instruments that any driver could have wished for.

Left **This rotary engine in RX-7 was fitted with an Elford Engineering turbo conversion. However, the initial emergence of the fuel-thirsty rotary engine coincided with the fuel crisis of the early 1970s, and drove customers away from showrooms, leaving Mazda to store its unsold cars.**

Power-assisted steering and four-wheel disk brakes allowed the car's performance to be exploited to the full. The car looked good and was very competent.

***Above* Turbo power by courtesy of Elford, turned Mazda's rather dowdy 929 coupé into something altogether more interesting and individual – a Q-car in the finest tradition.**

Mercedes-Benz 300SL

6-cylinder | 195 bhp | 140 mph | 1950s

The gull-winged Mercedes-Benz 300SL stood for 3-liter Sport-Leicht, or light sports car. The engine was big and heavy so the rest of the car had to be as light as possible, to which end the competition department designed a frighteningly complicated, stressed multi-tubular space frame, which weighed little more than 110 lb (50 kg).

To this they bolted coil spring front suspension, swinging arm rear suspension, and an aluminum coupé body with the famous gull-wing doors. To keep the hood line low the engine was canted through 40 degrees and offset to one side.

It was Mercedes' New York agent who suggested turning the 300SL into a road car and when the company expressed doubts he eased their uncertainty with a firm order for 1,000 cars! Production engines pioneered the use of Bosch direct fuel injection.

Although the road car was substantially heavier than the racer, and slower, it was still streets ahead of any opposition. On the lowest of three optional axle ratios it would reach 60 mph (96 kph) in under 7.5 sec. and 100 mph (160 kph) in less than 15 sec.

The chassis was a mixture of good and bad; the huge aluminum-finned drum brakes, with servo assistance, were exceptional, but the swinging arm rear suspension gave the sort of high-speed oversteer that only a really gifted driver could cope with.

Right **Wearing its three-pointed star with pride. Supersleek styling and hand-built craftsmanship combine to make the 300SL look right from every angle.**

Left **Probably the most famous Mercedes of all time, the gull-wing 300SL is race-bred. 140 mph (224 kph) and stunning acceleration are combined with somewhat treacherous handling — not a car for the faint-hearted.**

Mercedes-Benz 380SEC

V8 | 204 bhp | 133 mph (214 kph) | 1980s

The Mercedes-Benz 380SEC was a superb fast four-seater coupé, able to carry four adults.

Its 234 ci (3,839 cc) V8 engine produced 225 lb ft of torque at 4,000 rpm. Its 0 to 62 mph (0 to

***Right* With the elegant 380SEC coupé, Mercedes Benz showed that it was possible to combine performance with style and still retain a strong marque identity – even though the functional simplicity which characterized the marque was interpreted by some as being bland.**

100 kph) time was very good, at 6.4 sec., but the overall fuel consumption was only average, at 22 mpg.

In appearance the 380SEC rivals the very good looks of the BMW 635CSi and was without doubt one of the most handsome Mercedes-Benz cars.

In fact the 380SEC was so good looking that anyone thinking about adding any of the increasingly prevalent aftermarket body styling appendages would have thought carefully before changing the standard car's beautifully balanced appearance.

Mercedes-Benz 500SEL

V8 | 231 bhp | 140 mph (225 kph) | 1980s

The 500SEL was slightly cheaper than the 380SEC and used a 303 ci (4,973 cc) V8 engine. Fuel consumption for the 500SEL was slightly better than for the 380SEC, at an average of 23.6 mpg. For a large car it was surprisingly easy to drive.

It appeared to shrink as the miles flew past, allowing the driver to place the car very accurately on corners.

Brakes were disks all round, those at the front were internally ventilated and ABS came as standard. The transmission was the excellent Mercedes-Benz 4-speed automatic and the differential was a limited-slip unit. Suspension was independent all round, with the front having anti-dive characteristics and the rear incorporating anti-squat control. The seats were anatomically correct in design and very comfortable.

Below **The Mercedes-Benz 500SEL was a big car by any standards, but 231 bhp and exceptional aerodynamics added up to 140 mph (225 kph) performance and surprisingly good fuel economy.**

Mercedes-Benz 500SL

V8 | 231 bhp | 137 mph (220 kph) | 1980s

The 300SL was a classic example of the Mercedes-Benz model of vehicle and the "SL" label had come to be applied to many fine sporty convertibles from Mercedes, a tradition which the 500SL continued.

Below **With increasingly restrictive legislation demanding even better crash protection, particularly for the American market, the drophead sports car came close to extinction in the 1970s but, happily, cars like the 500SL show that it has not been stamped out yet.**

The all-alloy 303 ci (4,973 cc) V8 engine uses fuel injection, transistorized ignition, and hydraulic tappets and produces 231 bhp at 5,000 rpm.

A 4-speed automatic gearbox was fitted, as was a limited slip differential. Suspension was the same as on the 500SEL model, and the braking system was also similar.

The 500SL's top speed was 137 mph (220 kph) and its 0 to 62 mph (0 to 100 kph) time was 7.6 sec. Fuel economy was an average of 26.5 mpg.

Mercedes-Benz 600 Landaulette

V8 | 250 bhp | 130 mph (209 kph) | 1960s–70s

The most famous pre-war Mercedes-Benz had been the massive Grosser model, and by the 1960s the time seemed ripe for a new version of such a car to be built. The huge Type 600, which remained in production for nearly 20 years, was an excellent way to fill such a need.

It stood alone as the largest-ever Mercedes-Benz car to be sold in the post-war period, and at the time seemed unlikely ever to be displaced from that pinnacle – a fully specified example could weigh more than 6,000 pounds (2,721 kg).

It had a unit-construction body shell, with a choice of long or large wheelbases, three or four windows per side, and up to seven seats.

Below **In extra-long wheelbase form, the Mercedes-Benz 600 was an imposing, almost improbable, sight. This Landaulete has a mere seven seats, but is so large that it really needs two parking spaces to accommodate it.**

Power came from a 250 bhp V8 engine, the first-ever production V8 from Mercedes-Benz, and automatic transmission was standard. There was hydraulic power-assistance for almost everything, the suspension was self-leveling, and there were disk brakes all round.

For the first two years, every 600 built had closed coachwork, but in 1965 a state Landaulette was also put on offer on the shorter of the two wheelbases. The four standard doors and the rear quarter windows were retained, as was a metal roof over the front-seat occupants, while there was a bulky fold-down soft-top at the rear.

Above **Room enough, surely, for any potentate and his entourage this Mercedes-Benz 600 Landaulette has six passenger doors, and three rows of seats.**

The charm of the extremely expensive 600 model was not only that it had a superb soft ride, impressive stability, and a top speed of around 130 mph (209 kph), but it was also a machine which could be, and usually was, driven very slowly indeed, on ceremonial occasions. Trim and furnishings – acres of wood, leather, and related fittings – were all the highest possible standard and there was a great deal of passenger space in the rear seat.

Mercedes-Benz S and SS

8-cylinder | 200–300 bhp | 140 mph (225 kph) | 1920s

Blower Bentleys have superchargers that run constantly. Mercedes-Benz chose a different approach for its equally massive S and SS sports cars in the 1920s. The huge 6-cylinder engine normally ran unblown, but with an extra shove on the throttle pedal, the driver could cut in the supercharger for straight-line sprints.

Designed by Ferdinand Porsche, the S appeared in 1927 with a 6.8-liter, overhead cam engine and the blower standing vertically at the front of the block. Soon afterwards the more powerful SS took the power up to 200 hp from 7.1 liters.

Usually seen with low, tourer bodywork, their immense size and chromed triple exhaust pipes poking through the hood struck awe into bystanders.

But what makes the enthusiast's heart beat wildly today is the sight (or sound) of the short-chassis SSK. With the pedal down and the enlarged blower engaged, an SSK pumped out

Above **Mercedes-Benz S.**

Right **In a shortened, more powerful form, Mercedes boasted 225 bhp from the SSK, of which they built only around 30.**

225 hp through its skinny tires, enough to carry it to 140 mph (225 kph) and beyond. Production figures only ran to the low thirties, putting this fearsome machine into the top rank of collectable motors.

Mercedes-Benz did build an even faster version, a works race-entry known as the SSKL (L for leicht, or light). A big blower, drilled chassis, and 300 horsepower made this the ultimate development of the S series. Only a handful was built, but unscrupulous dealers are thought to have converted other models to this specification more recently.

Mercer 35R Raceabout

4-cylinder | 55 bhp | 70 mph (113 kph) | 1910s

The Mercer was perhaps the first car that could genuinely be called a "sports car."

It had a 4-cylinder 55 hp T-head engine, two bucket seats, a handbrake, gearshift, four wheels, and not much else. Even the throttle pedal was outboard of the driver's seat.

The bodywork involved no more than a low hood, long rakish wings, and a massive bolster fuel tank on the tail, surmounted by two spare tires on detachable rims.

The steering was light and precise, the gearshift incredibly slick, and the road holding was a tail-sliding delight. The brakes were terrible, but that never seemed to stop the Mercer driver from using his car's performance to the full. It was perhaps the first real sports car.

Below **Lightweight (and almost nonexistent!) bodywork coupled with a 4.9-liter engine, gave the Mercer Raceabout sparkling acceleration and a 70 mph (112 kph) top speed. Creature comforts are few and far between, and stopping the car is a bit of a problem. Available in only the most garish colors, the Raceabout is definitely the car to be seen in.**

MG Midget

4-cylinder | 46–66 bhp | 86–101 mph | 1960s–70s

Above **The MG Midget was also built in "badge engineered" form as an Austin-Healey Sprite. The only visual difference was to the badging and decoration of the car – this was an MG, as the familiar octagonal-shaped badge shows.**

The basic engineering of the Midget was of a Healey family design.

The car was small, and had a somewhat tight-fitting cockpit and rather basic trim and instrumentation. It had a sturdily engineered monocoque body/chassis unit with, in its original guise, a rear axle slung on cantilever leaf springs and radius arms.

To keep up with its competition, the Midget was regularly given more powerful and free-revving engines. There was a 1.1-liter engine for 1963, a more powerful version of the same in 1964, and a very tuneable1.3-liter enlargement of it for 1967. Then, for 1975, came the ultimate insult, as far as MG fans were concerned – the Midget inherited the 91 ci (1,493 cc) engine of the Triumph Spitfire!

The last Midget was the only derivative which could reach a genuine 100 mph (160 kph), and was still a nippy and enthusiastic little sports car.

Mitsubishi 3000GT VR-4

V6 | 296 bhp | 155 mph (249 kph) | 1990s

Mitsubishi unveiled its 3000GT VR-4 in 1991. What Mitsubishi set out to do was to cram every bit of high-tech gadgetry available into its 2+2 3000GT, which made it one of the most advanced sports cars on the planet during the 1990s. The car went from 0 to 60 mph (0 to 96 kph) in about 5 sec.

It had a twin-turbo, twin-intercooled V6, full-time all-wheel drive, speed-sensitive four-wheel steering, a computer-controlled suspension, a speed-activated system of aerodynamic aids, and an exhaust system that could be tuned for either power or quiet via a switch in the cockpit.

Each bank of three cylinders got its own turbocharger and intercooler. Fuel was fed to the V6 through a computer controlled multipoint injection system.

Mitsubishi engineers added a drive line to the rear wheels for four-wheel

Above **A cut-away drawing reveals the technical wonders of the 3000GT VR-4.**

drive. To move the engine's torque around the chassis, Mitsubishi developed an all-wheel-drive system that, under normal conditions, split the power so that 45% went to the front wheels and 55% to the rear.

The car had a 5-speed manual gearbox. And unlike some four-wheel-steering systems, the one on the 3000GT was designed strictly to aid high-speed cornering. When the steering wheel was turned, a computer monitored the speed of the car, how fast the steering wheel was being turned, and the lateral force exerted on the toe-control member of the rear suspension. The system allowed rear wheels to be turned up to 1.5 degrees in the same direction as the front wheels, making corner manoeuvres sharper.

Stopping power on the VR-4 was considerable, thanks to huge disk brakes at all four wheels.

Inside, the 3000GT VR-4 was surprisingly roomy at the front, with high-grade appointments throughout the cabin.

Above **Mitsubishi's 3000GT, in the foreground, was a direct descendant of the HSR-II prototype.**

Morgan Plus 8

V8 | 151–190 bhp | 125 mph (201 kph) | 1960s

The Morgan Plus 8 was born in 1968. It used the sliding pillar type of independent front suspension with which every Morgan had been fitted since the first tricycle had taken to the road in 1910.

Development, and change, came slowly at Morgan; but in the first 20 years the engine was gradually made more powerful, the entire gearbox was changed on two occasions, a rack-and-pinion steering installation became available, and the whole car became wider to match the latest tires and wheels. The car had a very hard ride.

Below **Except that wheels have become fatter, the front end style of the Plus 8, complete with podded headlamps and a traditional type of chrome bumper, is just the same as it was in the mid 1950s.**

Morris Minor

4-cylinder | 27–48 bhp | 60–75 mph (97–121 kph) | 1950s–60s

Above **Alec Issigonis's Morris Minor was launched in 1948 and built, with several different sizes of engine, until 1971; there was always a convertible in the range. All had a great deal of character, but of course, not much performance.**

With its rounded and altogether unmistakable looks, the Minor was launched in October 1948 and remained on sale, in a whole variety of guises, for the next 23 years.

The old side-valve engine gave way to a new overhead-valve BMC unit in the early 1950s and this, in more and yet more powerful guise, was retained until the end of production in 1971; a 1.1-liter unit was fitted from late 1962.

The vee-screen gave way to a curved one-piece screen in 1956, at which point the enlarged engine gave it the "Minor 1000" title, but after the end of the 1950s little further development took place, and the car gradually died away. Even so, something of a Morris Minor "cult" grew up as interest in classic cars developed in the 1970s.

Nissan 300ZX Turbo

V6 | 300 bhp | 155 mph (249 kph) | 1990s

If history and pedigrees count for anything, Nissan's 300ZX was the most venerable of all Japanese sports cars. When the model was introduced in 1970 as the 240Z, it was a revolutionary idea – a sports car that performed well, looked great, and was as reliable as any car.

From that solid base, Nissan gradually improved on the Z car. Although there were a few hiccups, each successive Z was an improved, more sophisticated car, representing Nissan's perception of what drivers really wanted.

When more power was the order of the day, Nissan added turbo charging; when luxury touches were required,

Above **A rare Nissan 300ZX without the T-bar roof panels.**

the 280ZX was rolled out; when a rudimentary back seat was needed, a 2+2 model was added; when open-air driving became popular, a T-bar roof with removable panels was incorporated into the design.

By the late 1980s, when the perception was that drivers wanted a boulevard car that looked good but was not too aggressive, the initial 300ZX was the perfect answer.

The 2.96-liter V6 engine was boosted by a Garrett turbocharger. Electronic fuel injection and ignition were fitted to aid the splendid flexibility which is one of the V6's best features.

Road holding was uniformly good and the car had no particularly bad habits so long as its understeering tendency was kept in mind. The brakes, disks all round, were well up to the task of stopping the 300ZX Turbo at all the speeds.

Below **Nissan started with a blank sheet of paper when it set out to design the new generation Z car.**

Nissan Silvia Turbo

4-cylinder | 135 bhp | 126 mph (203 kph) | 1980s

The smaller sister car to the 300ZX Turbo was much more in the cast of a European sports coupé. This was the Nissan Silvia Turbo ZX.

It used a turbocharged 1.8-liter overhead-cam engine and an excellent fuel economy figure of 44 mpg at 65 mph (105 kph).

For lazy drivers, the Silvia could be ordered with a 3-speed plus overdrive top automatic gearbox but the manual 5-speed box was really too

good to pass up. With all-round independent suspension, rack-and-pinion steering, and four-wheel disk brakes, the chassis was well equipped to handle the engine's fine performance.

The Silvia was better balanced than the 300ZX Turbo, and had an outstanding ability to cover the miles fast and safely. Its styling was perhaps just a little too bland, even anonymous, to be a head-turner, but it is clean and aerodynamically stable.

The car was a pleasure to drive when carrying just two people and their luggage but as a 2+2 it had only limited use for four adults.

Above **The Nissan 300ZX Turbo started with the 240Z.**

Left **The Nissan Silvia Turbo was Nissan's alternative performance car.**

Opel Monza GSE

6-cylinder | 180 bhp | 133 mph (214 kph) | 1980s

The Opel Monza GSE was a striking looking car that was very well made, most fully equipped, had really excellent road manners, and was safe and fast at the same time. The hatchback GSE was able to carry four adults and their entire luggage over long distances in great comfort.

It was powered by a 3-liter straight-six overhead cam engine developing an impressive 183 lb ft of torque at 4,200 rpm. Bosch LE-electronic fuel injection and electronic ignition were fitted and gave impressive smoothness and power.

Either a 5-speed manual or a 4-speed automatic transmission

could be ordered with this engine and both options allowed the full performance potential to be exploited.

Light alloy wheels helped brake-cooling and enhanced the overall appearance of this car. Its acceleration times were good, for the 0 to 62 mph (0 to 100 kph) dash the manual took only 8.2 sec. and the auto-equipped car was just 2 sec. slower. The top speed was 133 mph (214 kph).

Four-wheel disk brakes were fitted and these were ventilated at the front and solid at the rear, assisted by a servo and regulated to prevent rear-wheel lock-up in emergency operation.

Below **The GSE's smooth fuel injected straight-six gave the big, well-equipped fastback coupé sparkling performance at remarkably low cost.**

Peugeot 205 Turbo 16

4-cylinder | 200 bhp | 130 mph (209 kph) | 1980s

When it was introduced in early 1983, the Peugeot 205 Turbo 16 had up-to-date styling, fine performance, and very good ride and handling. The car was met with great enthusiasm, and the company had to subsequently increase production rates of the car several times to meet demands.

The company also went rallying, but with a car that was even more

outstanding: the 205 turbocharged 16-valve engined, four-wheel-drive machine. They won rallies, including the 1985 Monte Carlo. To be able to enter this extraordinary car in international competition Peugeot had to build at least 200 customer examples, which they did, and sold every available car.

The car's 108 ci (1,775 cc) engine used four valves per cylinder and is blown by an intercooled German-made KKK turbocharger. If it were not for the four-wheel-drive the car would be difficult to control.

Driving this competition-inspired vehicle called for firm and positive action on the part of the driver – the Peugeot had to be made to obey directions and this was no car to pussyfoot around; it had to be dominated. There was little luggage room, it was noisy, and could be tiring, but it was fun to drive far.

Left **The Peugeot 205 Turbo 16 in works competition guise ended the Audi Quattro's near monopoly of rally wins in 1984 and the model also became available as a rather special road car.**

Pontiac Fiero

V6 | 140 bhp | 127 mph (204 kph) | 1980s

Apart from the evergreen Corvette and a few low volume specials and replicas, America hasn't had much to offer by way of sports cars since the mid 1960s. Yet with the Pontiac Fiero, introduced in 1983, the least likely GM division produced a car that was at once innovative, admirably functional, and quite capable of taking on many of Europe's and Japan's best sports cars both in terms of speed and style.

It was based on a steel inner shell, with a prominent backbone, on which all the mechanical bits append to make a completely drivable rolling chassis. This was then clothed in an all-plastic outer shell, attached to machined locating pads to ensure a perfect panel fit.

It was clever, pretty, and a frustrating disappointment to real enthusiasts, because Pontiac had failed to go the whole way. They had

launched the mid-engined coupé as a sporty commuter car rather than a sports car, with an old and asthmatic 91 bhp, 2.5-liter, 4-cylinder, cast-iron engine that left it barely able to get out of its own way. With this engine, the dynamic abilities of the Fiero were wasted.

The cure came within a year, as Pontiac gave the Fiero a rather more exciting fuel-injected 2.8-liter V6. In the GT version, it offered almost 50% more power, and an equally useful 170 lb ft of torque. In GT form, its top speed went up to over 125 mph (200 kph) and the 0 to 60 mph (0 to 96 kph) time came down to a far more respectable 8 sec.

Left and Above **Mid-engined and plastic-bodied, with superb handling, the Fiero embodied all the latest in sports car technology, but threw it all away with its outdated underpowered engine.**

Suspension was achieved by employing unequal length wishbones at the front and a Chapman strut arrangement with lower wishbones at the rear, with coil springs, telescopic dampers, and anti-roll bars at both ends. The steering, while feeling sharp enough, was slower than the sports car norm but there was enough built-in understeer to make the Fiero inherently very safe for all but the clumsiest driver.

Porsche 356 Cabriolet

Flat-4 | 40–90 bhp | 85–111 mph (137–179 kph) | 1950s–60s

There were four distinctly different types of 356 – the original vee-screen cars built up until 1955, the Type 356A of 1955–1959, the 356Bs of 1959–1963, with raised headlamps and bumpers, and the final Type 356C of 1963–1965, which had disk brakes.

In that time the engines were gradually but persistently enlarged, with many different sizes being offered along the way. The majority of surviving 356s, however, now seem to have the 96 ci (1582 cc) engine, either in 75 bhp or 90 bhp guise.

The platform chassis had its engine in the tail, driving forward to a gearbox/transaxle, while there was independent suspension at front and rear. Most of the car's weight was in the rear with lots of tail-out oversteer.

All these Porsches had remarkably efficient body styles and because they were also high-geared it was possible to cruise along very quickly for hours on end. Most early cars were fixed head/fastback coupés, but the Cabriolet, whose top tucked neatly away when furled, and had a very small rear window, was also popular.

Right **A historic Porsche – actually the first car produced by the fledgling firm in 1948. This was a very starkly detailed machine, which was almost entirely VW Beetle under the smooth skin.**

Right **Over the years the Porsche 356 was produced with several different body types. The Cabriolet was on sale from the start, and was later joined by a hardtop version using the same body pressings. The engine was in the tail and much of the chassis engineering derived from that of the VW Beetle.**

Porsche 356 Speedster

4-cylinder | N/A | 120 mph (193 kph) | 1950s

The 356 was the first car to bear the Porsche name, though not the first to bear the stamp of Porsche's genius. Nominally, it was the 356th design of the Porsche design bureau, set up in Stuttgart in 1930. This was somewhat misleading, because Dr Porsche started with Project 7, to avoid his first customers thinking he was inexperienced.

The 356 had an open body and mid-engine, but all subsequent cars were rear-engined, to offer adequate cockpit space.

In August 1948, Porsche completed the first 356 coupé and started small-scale production, launching the car officially early in 1949. The 356 developed rapidly, with the first capacity increase, to 78 ci (1286 cc), in 1951, followed by another, to 96 ci (1582 cc), in 1955 – this being the basic capacity until the last of the 356s in 1965. There was only one major body change, in 1959, with a larger windscreen and slight raising of the bumpers and headlamps, but there were many variations on the 356 theme, both open and closed.

The most distinctive of all 356s was the stunning but short-lived Speedster. The Speedster, a sparsely equipped, lightweight, open 356 with a chopped down windscreen, clearly evoked the racing image. It had near racing performance, with a 120 mph (193 kph) top speed, 10 sec 0 to 60 mph (0 to 96 kph) times, and superb

Right **The mark of genius. Light and aerodynamic, the short-lived Porsche 356 Speedster uses the classic rear-engine configuration still used by Porsche to this day.**

brakes and transmission. Whatever people said about Porsche handling, very few cars could pass one.

It was introduced in 1954 after American importer Max Hoffman had bought a racing Spider and mooted the look alike to Porsche. It was dropped when Porsche realized they could not possibly make money on such a basic car.

Above **The dash incorporates only a speedometer, rev counter, and an oil temperature gauge housed in an attractive Italianesque pod.**

Porsche 911 Turbo

Flat 6 | 300 bhp | 160 mph (257 kph) | 1980s

The Porsche 911 Turbo model was known as the 930 in North America. It was introduced way back in late 1964, though progressive improvements over the years have upped the 911's power output.

Any doubts, however, as to whether the 911 chassis could really cope with 300 bhp had already been allayed by even more powerful racing derivatives and the 911 Turbo had all the necessary refinements to enable it

to handle this massive amount of power with ease.

Like all the 911 models, the Turbo retained some aspects from its past, in particular, its instruments and heating. As with all air-cooled cars the efficiency of the 911's heating and ventilation system varied with the engine speed.

The instrumentation of the 911 series lagged behind cars of its time; with the exception of the speedo and rev-counter, the dials and switches were scattered about the dash.

The 911 Turbo may have been flawed, but its virtues outweighed its faults by a very wide margin.

An acceleration from 0 to 62 mph (0 to 100 kph) in just 5.4 sec., fuel consumption of 23.9 mpg at a steady 75 mph (120 kph), together with the car's incredibly stable resale value really put the Porsche 911 Turbo into a very special category.

Left and Above **Although introduced as long ago as 1975 and based on a series which dates from 1964, the Porsche 911 Turbo was the fastest accelerating car of any in series production during the 1980s. Its understated, almost mundane looks belie truly stunning performance. Porsche pioneered the use of aerodynamic wings on road cars and, with a car as quick as the Turbo, they were not simply for show.**

Porsche 944 Turbo

4- cylinder | 220 bhp | 152 mph (245 kph) | 1980s

Early in 1985, Porsche announced their latest high-performance car, the long awaited 944 Turbo model. The body styling was little different from the non-turbo 944.

The 944 Turbo's front-mounted 4-cylinder engine produced a claimed 220 bhp, enough to propel the car to a top speed of 152 mph (245 kph) and cover the 0 to 62 mph (0 to 100 kph) run in 6.3 sec.

The car's turbocharger was water cooled, and both an intercooler and engine oil cooler were fitted, displaying again the attention to detail which was a recurring feature of the Porsche way of doing things.

The improvement in performance over the standard 944 could be quickly gauged from a top speed of 137 mph (220 kph), 0 to 62 mph (0 to

100 kph) in 8.4 sec. for the "ordinary" car, and very similar fuel economy.

On each model of 944 certain common features remained: low-drag aerodynamics; a good balance between performance and fuel economy; high top speed; good handling, steering, and braking.

Below **When the Porsche 944 was introduced in 1982 it helped bridge a gap between the relatively low-powered 924 and the quick but very expensive 928, which was really more of an executive express. It combined what were essentially the 924's mass-produced shell, with some modification, and half the 928's V8 engine. This 944 Turbo (*above*), introduced in 1985, took the car a step further toward 911 performance.**

Porsche 959

Flat-6 | 400 bhp | 190 mph (306 kph) | 1980s

Porsche has a reputation for unconventional but effective engineering. In 1981 the Stuttgart firm amazed the automotive world with a prototype called the 959. Without abandoning the theme of the longest-running sports car of all, the 911, the company's Weissach development center, produced a fast and technically sophisticated car, an all-wheel drive, 450 bhp sensation capable of vitually 200 mph (322 kph). It was the first of a new generation of super cars capable of doubling the magic "ton." The 959 achieved another double when it won the Paris-Dakar rally across Africa in 1986 and 1987 and raced at Le Mans a year later – the first "road-going" (albeit highly modified) sports car to do so for 20 years. New racing rules later made it redundant as a competition car.

The mid-engined 959 used a refinement of Porsche's existing 935/936 racing engines. It was a twin turbocharged 2.85-liter flat-six unit with an intercooler. The cylinder heads featured four valves per cylinder and water cooling. The engine block itself, however, was air cooled.

The engine's power output was rated at a minimum of 400 bhp and the 959 accelerated from 0 to 62 mph (0 to 100 kph) in 4.9 sec.

Its body was made of Kevlar, a material more usually found in the bodies of endurance racing or Grand Prix cars. The Kevlar body was mounted on a galvanized steel chassis that carried racing type suspension with dual wishbones at front and rear. Twin dampers were fitted all round, and those at the front incorporated dual springs. The driver could adjust both ride height and spring rates from the cockpit.

Porsche fit magnesium wheels with hollow spokes to the car, which allowed the wheel and tire assembly to be monitored constantly for punctures or structural failure. The four-wheel-drive system was electronically controlled, to give optimum traction to the car by varying the front-to-rear drive balance automatically at all speeds.

Left and Below **The 959 was the fastest and most expensive car Porsche made for the road in the 1980s. Although the shape was clearly related to the 911, the car owed more allegiance to the 935 racers.**

Renault Alpine GTA V6 Turbo

V6 | 200 bhp | 152 mph (245 kph) | N/A

The GTA is rear-engined, with the engine positioned behind the rear axle-line of the neat backbone chassis. The glass-fiber-reinforced monocoque body shell is a reasonably spacious 2+2 seater and offers very good standards of interior trim and equipment.

The most potent engine is the 2.5-liter V6 with a single overhead camshaft on each cylinder bank and a Garrett turbocharger. This smooth, free-revving unit, coupled to a 5-speed gearbox, gives the GTA quite exceptional performance, with a 0 to 60 mph (0 to 96 kph) acceleration time of less than 6 sec.

Its handling is generally, but not always, taut and predictable, with coil springs and double wishbone suspension all round and very precise rack-and-pinion steering.

Below **The Turbo version of the Alpine, Renault's venture into sports cars, using the Alpine name made famous in motor racing. The Alpine GTA proves that even the mass production manufacturers can produce classic sports cars.**

Renault Floride & Caravelle

4-cylinder | 40–55 bhp | 89 mph (143 kph) | 1960s

The original Floride of 1959 (called Caravelle in the American market) was built on the same underpan as the Dauphine Gordini but had attractive and more angular styling. At first this car used the 52 ci (845 cc) engine, but in 1962 the design was rejigged. By the end of 1963 that new engine had already been enlarged, to 68 ci (1108cc), and the development process was complete.

Because the Floride/Caravelle was a model with more glitz than performance, and with rather bright and glossy trim, it tended to be written off by the purists, and was even given that dreadful nickname of a "hairdresser's car." Yet it sold well, especially in France and other Mediterranean countries.

Above **Although it isn't easy to produce a graceful style on a rear-engined car, Renault succeeded with the Caravelle. A bigger engine and more performance would have made it a more desirable car.**

Rolls Royce Phantom II

6-cylinder | N/A | 90 mph (145 kph) | 1930s

The Rolls-Royce chassis of the day was conventional, but exquisitely built. Engines were low-revving 6-cylinder units, brake operation was assisted by the Hispano-type of mechanical servo, while front and rear suspension was by half-elliptic springs and lever-arm dampers.

Rolls-Royce completed the chassis, which then could be driven to the chosen coachbuilder. With the chassis came the company's famous Palladian-style radiator, as well as the scuttle-firewall and instrument panel, both retained at the company's insistence.

Every type of body was, of course, married to these chassis, ranging from staid and upright limousines with divisions, through rakish "owner-driver" sports sedans, to any number of startlingly elegant open-top machines.

Below **Coachbuilders produced a huge variety of styles for the 20/25 chassis, not only open-top but sedans and limousines. All of them were graceful and dignified.**

Above **Look carefully at this exquisitely bodied Phantom II, and it becomes obvious why every gentleman wanted to be seen in one in the 1930s. All the details are typical of a twenties/thirties coachbuilt style, including the mounting of the spare wheel on the side of the body.**

Invariably the shell was built up around a skeleton of seasoned hard wood, and on many occasions the hand-beaten paneling would be in rust-proof light alloy. Wings and running boards, however, were usually in steel, which was better able to withstand the assault of stones from the road surface.

The variety, and the sheer inventiveness of the coachbuilders concerned, was a joy. Some cars had one or even two exposed spare wheels tucked into recesses in front wings. Some had integral luggage boots but others used demountable containers; some used wheel disks, while others left their wire-spoke wheels exposed. All had massive headlamps at each side of the grille, that famous radiator and the "Spirit of Ecstasy" on the prow.

Rolls Royce Phantom V & VI State Landaulette

V8 | N/A | 100 to 105 mph (161 to 169 kph) | 1960s onward

In 1959 the long-running Silver Wraith was retired, and immediately replaced by the massive yet elegant Phantom V. Every Phantom V, and Phantom VI's which took over in 1968, was coachbuilt.

The Phantom was too large a car for a full convertible body style to be

thought practical or elegant. Rolls-Royce therefore evolved a half-and-half design in which most of the shell remained closed but the rear section could be opened up. This long-established style was, and is, known as a Landaulette.

The first Landaulette derivative of the Phantom V was built by Mulliner Park Ward in 1962, and delivered to H.M. the Queen Mother. For this car, there was a fold-down soft-top above and behind the rear seats only.

Later, Rolls-Royce offered a slightly different version, in which the folding roof was longer, and was fixed to the top of the division between front and rear compartments. This car also had an electrically operated rear seat which could, if needed, be raised by 3.5 in. (8.9 cm). They were, need it be said, extremely exclusive machines.

All the statistics were large and impressive and like all such "ceremonial" Rolls-Royces, they were cars in which to be seen, and in which to be driven around town, or from public engagement and board meeting to the town house or country dwelling, rather than as regular, long-distance, business transport.

The Phantom V became the Phantom VI in 1968, with minor changes. The enlarged 6.75-liter engine soon followed, and in 1978 the chassis was given the high-tech hydraulic circuitry of the Silver Shadow.

Left **Although it had graceful and dignified lines, the Phantom VI was a very large car, nearly 20 ft (6.1 meters) long . There was ample room for five in the rear compartment, though only those sitting in the rear seat itself got the benefit of the open-top when it was furled.**

Saab 900 Cabriolet Turbo 16

4-cylinder | 175 bhp | 125 mph (201 kph) | 1980s

Above **Except that there was the inevitable visibility blind spot in the rear quarters, the 900 Cabriolet Turbo 16 was a thoroughly practical convertible car: customers seemed to think it was well worth the premium price asked.**

A new generation of passenger cars, the Saab 99 was introduced in 1967, and this was greatly refined, as the longer wheel base Saab 900, in 1978.

The prototype Cabriolet was shown in 1983 and it took well over two years before it could go on sale. Because Saab saw the Cabriolet as its most prestigious model, it was not only given full air-conditioning for sale in the United States, and a power-operated soft-top, but also had an ingeniously arranged glass rear window which dropped into a pouch ahead of the boot when the soft-top was furled.

The engine chosen was the very latest twin-overhead-camshaft turbocharged 2-liter unit, which had four valves per cylinder.

Saab 9000 T

4-cylinder | 175 bhp | 130 mph (209 kph) | 1980s

The Saab 9000 is a sober, functional shape, with enormous interior space, a well laid out set of instruments and controls to aid the driver at his task, and, under the bonnet, all the latest sophistication to be expected of a fourth generation turbo car.

It has stability at speed, it has excellent road holding and ride qualities, and it shows all the effort put into it after ten years of painstaking research and development. The 121 ci (1,985 cc) 4-cylinder engine has four valves per cylinder and double overhead cams and intercooled turbo unit.

Below **The attractive Saab 9000 sedan featured the first all-new bodyshape from the Swedish manufacturer, and much new engineering under the skin further enhanced Saab's considerable reputation for solid sporting performance.**

Toyota Celica Supra

6-cylinder | 168 bhp | 136 mph (219 kph) | 1980s

The Celica Supra 2.8i was updated in the 1980s, with the suspension department displaying the first results of the successful Toyota/ Lotus partnership.

The engine is a classic with a 168 ci (2,759 cc) twin-cam straight-six driving the rear axle via a 5-speed gearbox. Brakes are ventilated disks at both front and rear, with servo-assistance. Steering is by an excellent power assisted rack-and-pinion system.

These features, plus the use of 225/60VR 14 tires and the Lotus-inspired changes to the chassis, make for a car that although lacking the power of its turbocharged rivals, can happily keep up during high-speed cross-country travel.

The engine and transmission are a delight. Not having turbo-boost means not having any turbo-lag, and the spread of engine power is such that the Supra is more relaxing to drive in all conditions of road, traffic, and weather than most.

Right **Celica Supra offers proof that one of the world's biggest motor manufacturer still believes in performance.**

TOYOTA

Toyota Corolla GT

4-cylinder | 122 bhp | 125 mph (201 kph) | 1980s

The Toyota Corolla GT is a sister car to the MR2.

Designed to compete with the fast hatchback models that flooded the market during the 1980s, the Corolla GT has the same superb 16-valve 97 ci (1,587 cc) engine as the MR2, but front-mounted and driving the front wheels. Unlike the MR2 it is a genuine four-seater.

Disk brakes all-round, rack-and-pinion steering, alloy wheels and 186/60HR 14 tires, and all-independent suspension with gas-filled dampers give the GT performance that enables it to challenge the mighty VW Golf GTi.

Because of the higher seating position and the better visibility it is preferable to the MR2. Its low price makes it particularly competitive.

Below **Sharing the same basic powertrain as the new mid-engined MR2, the 125 mph (201 kph) Corolla GT finally lays to rest Toyota's formerly well-deserved image as a maker of very ordinary motorcars.**

Toyota MR2

4-cylinder | 122 bhp | 125 mph (201 kph) | 1980s

The mid-engined Toyota MR2 was announced to the British press in Portugal in January 1985. The car uses a 4-cylinder engine, the twin-cam 4AGE unit of 97 ci (1,587 cc) capacity and drives through a 5-speed gearbox and transaxle arrangement.

The MR2 accelerates to 60 mph (96 kph) from rest in 8.1 sec. At a constant 75 mph (121 kph) it has excellent 36.7 mpg fuel economy. The electronic fuel-injection system gives instant starting, fast cold drive-away, and superb response at all engine speeds, so good as to make the 5-speed gearbox almost redundant!

Steering, brakes, stability at speed, driving position, controls, finish, were all considered good. On the flip side, the fuel tank is too small at only nine gallons and the interiors are a little cramped.

Below **The Japanese car giant Toyota brought the mass-produced mid-engined sports coupé to the world's attention again with their MR2.**

Triumph Stag

V8 | 145 bhp | 116 mph (187 kph) | 1970s

The Triumph Stag appeared in 1970 and started out as a one-off project car, a plaything cobbled up by Michelotti, in Turin, as a convertible version of the Triumph 2000 sedan. Even in that form, however, it was promising enough to be snapped up by Standard-Triumph for development as a production car.

For its production program Triumph decided to shorten the wheelbase, turn it into a close-coupled four-seater, and use a fuel-injected TR5 type of engine. The 6-cylinder engine was also discarded, in favor of the new 90-degree V8 Triumph unit and it became a 183 ci (2997 cc) unit.

In basic form, the Stag was sold as a convertible, with a fold-away soft-top. A very popular option, however, was a sturdy steel hard-top which included glass rear quarter windows and a glass backlight. The car made no sacrifices in comfort.

Above **The Stag was an appealing 2+2 seater package, with V8 engine, all independent suspension, and a choice of soft-top or hard-top body styles.**

Triumph TR2

4-cylinder | 90 bhp | 105 mph (169 kph) | 1950s

The TR part stands for Triumph Roadster, the name given to the first open-topped post-World War II Triumph, which appeared in 1946.

The car was well received and with a few months' refinement it re-emerged in mid 1953 as the TR2. The TR2 had its chassis under slung at the rear, an engine linered-down to just under 2 liters, gearbox and semi elliptic rear axle from the Vanguard, and front wishbone and coil spring suspension.

Its chunkily attractive slab-sided body also reflected low tooling costs and hurried development, but somehow it all clicked and the little Triumph was a major success – as a racing and rally car as well as just in the showroom.

***Above* Hurried into production, the lab-sided TR2 was not the sleekest new sports car around. However, what it lacked in looks it made up for in driveability on both road and track.**

Volkswagen Golf GTi Cabriolet

4-cylinder | 110 bhp | 113 mph (182 kph) | 1980s

The Golf went on sale in 1974, and there were no plans to make a convertible car at first. But a Cabriolet version of the monocoque Golf eventually surfaced in 1979. The new model was styled, developed, tooled, and put into production by the specialist coachbuilders, Karmann.

The Golf Cabriolet, if for no other reason, is important as the first of the modern generation of convertibles, almost all of which are conversions, or re-engineering projects, based on mass-produced sedans or hatchbacks.

Even though the Golf hatchback itself was completely redesigned in 1983, the original and much-loved Cabriolet style of 1979 carried on into the late 1980s.

Below **In 1979 VW virtually re-invented the soft-top car for family motoring, this being an extensively redesigned version of the famous hatchback; it was built, for VW, by the coachbuilders Karmann.**

Index

JAGUAR
SPORT
XJR 15
GOODYEAR
EAGLE